Mathematics of Casino Carnival Games

Mathematics of Casino Carnival Games

Mark Bollman

CRC Press
Taylor & Francis Group
Boca Raton London New York

CRC Press is an imprint of the
Taylor & Francis Group, an **informa** business
A CHAPMAN & HALL BOOK

First edition published 2021
by CRC Press
6000 Broken Sound Parkway NW, Suite 300, Boca Raton, FL 33487-2742

and by CRC Press
2 Park Square, Milton Park, Abingdon, Oxon, OX14 4RN

First issued in paperback 2022

**Visit the Taylor & Francis Web site at
http://www.taylorandfrancis.com**

**and the CRC Press Web site at
http://www.crcpress.com**

Library of Congress Cataloging-in-Publication Data
Names: Bollman, Mark, author.
Title: Mathematics of casino carnival games / Mark Bollman.
Description: First edition. | Boca Raton : Chapman & Hall, CRC Press, 2020.
 | Includes bibliographical references and index.
Identifiers: LCCN 2020017695 (print) | LCCN 2020017696 (ebook) | ISBN
 9780367348656 (hardback) | ISBN 9780429328527 (ebook)
Subjects: LCSH: Games of chance (Mathematics) | Gambling--Mathematics.
Classification: LCC QA271 .B64 2020 (print) | LCC QA271 (ebook) | DDC
 519.2/7--dc23
LC record available at https://lccn.loc.gov/2020017695
LC ebook record available at https://lccn.loc.gov/2020017696

ISBN 13: 978-0-367-33977-7 (pbk)
ISBN 13: 978-0-367-34865-6 (hbk)
ISBN 13: 978-0-429-32852-7 (ebk)

DOI: 10.1201/9780429328527

Typeset in LMRoman
by Nova Techset Private Limited, Bengaluru & Chennai, India

For my generation:
Laura, Dan, Anne, Monica 1, Eric, Monica 2, Kristine,
Laura, John, Amy, Chuck, Jif, Dave, Steven, and Robyn.

Contents

Preface

There are:

- 52 cards in a standard deck.
- 38 pockets on an American roulette wheel.
- 6 sides on a standard casino die.

These 96 game elements (and a few extras, here and there) have been combined over the years into a wide variety of casino table games. While the "big four" games of blackjack, baccarat, craps, and roulette have enduring popularity and a long history on casino floors, there are hundreds of other games played with these simple devices that have found (or hoped to find) a place—even if only for a short time—in legal and illegal casinos. These "carnival games" and the mathematics behind them are the focus of this book.

Why are they called carnival games? Some of these games, including chuck-a-luck and the Big Six wheel, have had a dual life in actual carnivals and in casinos. As carnival midway games, they were sometimes crooked games as operators sought to make as much money as possible during the limited season; in the casinos, their frequently high house advantage eliminates the need to run a dishonest table.

Carnival games range throughout time: some of these games have a long history, often with origins in illegal gaming establishments. Faro was one of the most popular card games during its heyday in the 19th and early 20th centuries before fading to extinction. Others have been proposed far more recently in an effort to be the next hot casino game. Some game designers have been motivated by the success of Caribbean Stud Poker, which was first devised in the 1980s and eventually sold for $30 million [45].

Acknowledgments

Once again, it has been a pleasure to work with the editorial team at Taylor & Francis in bringing this project forward. Saf Khan and Callum Fraser have been great editors and great colleagues. Vaishnavi Ganesh at Nova Techset has been a fine project manager moving the book through production. Thanks again to Robin Lloyd Starkes, serving for a third time as Production Editor and moving a book of mine through the production process.

Some of this material was class-tested in Math 257: Mathematics of the Gaming Industry, at Albion College in the spring of 2020. I received good feedback on several sections from Subeedei Barkhasbadi, Altanzul Davaa-Ochir, Kelsi Inman, Kurtis Sandford, Slone Schultz, Jared Simkins, and Samantha White.

The Python simulations of Twenty-Six in Chapter 5 were performed by students in my Computer Science 256: Practicum in Python Programming course at Albion in the spring of 2018. Thanks to Erris Canamusa, Justin Leeds, Claire Ostrowski, Josh Pemberton, Robert Petersen, Shawn Roberts, and Andrew Strzelecki for their contributions.

Spider craps (page 29) was developed by Albion student Jacob Engel during a summer research program in 2011. Funding for this project was provided by Albion's Foundation for Undergraduate Research, Scholarship, and Creative Activity (FURSCA).

This work was supported in part by a 2017 grant from the Hewlett-Mellon Fund for Faculty Development at Albion College, Albion, MI. This grant supported my travel to Las Vegas for three weeks of study at the Center for Gaming Research at the University of Nevada, Las Vegas. I am very grateful for the assistance of the staff at the Center during those three weeks and a return visit in the spring of 2019.

The Rocket Roulette image in Chapter 1 is reprinted from *Systems & Methods*, volume 9 (1975), published by Gamblers General Store. Permission was granted by Adam Pennell of GGS. Some material in Section 4.1, on Three Card Poker and Casino War, is reprinted from my previous book, *Basic Gambling Mathematics: The Numbers Behind the Neon*, © 2014, and is reproduced by permission of Taylor and Francis Group, LLC, a division of Informa plc. Permission was conveyed through Copyright Clearance Center, Inc.

Chapter 1

Mathematical Background

Before diving into carnival games, it will be useful to review the mathematics that is essential to analysis of games of chance. In this chapter, we shall do this through the lens of familiar casino games such as craps, roulette, and blackjack.

1.1 Elementary Probability

An *event*, in a probabilistic sense, is simply the outcome of some sort of random experiment. If A is an event, the *probability* of A is a function $P(A)$ that assigns a number to A, in the interval $0 \leqslant P(A) \leqslant 1$, as a measure of how likely A is to occur. Informally, we might define $P(A)$ as

$$P(A) = \frac{\text{Number of ways that } A \text{ can happen}}{\text{Number of ways that something can happen}}.$$

The set of all things that can happen in a given experiment, counted in the denominator, is called the *sample space* and often denoted by **S**. If we define a function $\#(A)$ that counts the number of elements in the event A, we can write

$$P(A) = \frac{\#(A)}{\#(\mathbf{S})}.$$

Counting the numerator and denominator of the expression for $P(A)$ is typically done either by appealing to pure mathematical reasoning, the *theoretical probability*, or by looking at real data, the *experimental* or *empirical probability*.

Example 1.1. In tossing a fair coin, the theoretical probability of Heads is ½: there is one way to throw Heads, and two ways for the coin to land.

If instead we were to toss a fair coin 500 times, achieving 256 Heads and 244 Tails, the experimental probability of Heads is $\dfrac{256}{500} = .512$. ∎

The connection between theoretical and experimental probability is described in a mathematical result called the *Law of Large Numbers*, or LLN for short.

Theorem 1.1. *(Law of Large Numbers) Suppose an event has theoretical probability p. If x is the number of times that the event occurs in a sequence of n trials, then as the number of trials n increases, the experimental probability x/n approaches p, or*

$$\lim_{n \to \infty} \frac{x}{n} = p.$$

Informally, the LLN states that, in the long run, things happen in an experiment the way that theory says that they do. What is meant by "in the long run" is not a fixed number of trials, but will vary depending on the experiment. For some experiments, $n = 500$ may be a large number, but for others—particularly if the theoretical probability is small—it may take far more trials before the experimental probabilities get closer to the theoretical probability.

Example 1.2. Consider the experiment of spinning an American roulette wheel with 38 pockets, and let A be the event that 00 is spun. The theoretical probability of A is $\frac{1}{38} \approx .0263$.

Computer simulation of 10 spins gives the experimental probability $P(A) = 0$, meaning that 00 never came up in 10 spins. After 100 spins, there were 7 00s, so the experimental probability is now .07. Simulating 1000 spins gives $P(A) = .021$, which is closer to $\frac{1}{38}$, but not as close to the theoretical probability as we saw with even 500 tosses of a coin.

10,000 simulated spins give 242 00s and an experimental probability of .0242, so the convergence of x/n to p is slow, but as n increases, we can be confident that the two values get acceptably close. ∎

What is meant by "acceptably close"? Formally, if $\epsilon > 0$ is any small positive number, the LLN states that, for a sufficiently large number of trials n, the difference between the theoretical probability p and the experimental probability x/n will have absolute value less than ϵ. However, "sufficiently large" is not precisely defined; it will certainly depend on ϵ, and is likely to depend on p.

The Law of Large Numbers does not require that independent events somehow "balance out," with short-term deficits of one outcome countered by surpluses. As an illustration, consider repeated tossing of a fair coin and suppose that the first 5 tosses are Heads. The experimental probability of Heads at this point is 100%, far from the theoretical probability of ½. However, if the next 200 tosses are merely split 50/50 between Heads and Tails, the experimental probability of Heads is now $\frac{105}{205} \approx .5122$, which is much closer to the theoretical value. This does not mean that the initial 5-tail deficiency has been made up.

The LLN is key to the success of the casino industry. While players may win small amounts from time to time, over the long term—and casinos are far better capitalized for the long run than any gambler can be—the casinos will extract essentially as much money as the laws of probability state they should.

The Addition and Multiplication Rules

Computing the probability of a more complicated event may be facilitated by breaking that event down into simpler events and carefully combining those probabilities.

Definition 1.1. Two events A and B are *mutually exclusive* if they cannot occur together; that is, if the occurrence of one implies that the probability of the other one occurring is 0.

Example 1.3. If we draw one card from a standard deck, the events $A = \{$The card is a club$\}$ and $B = \{$The card is a spade$\}$ are mutually exclusive, since each card has only 1 suit. The event $C = \{$The card is a black card$\}$ is the event that either A or B occurs.

We have

$$P(A) = P(B) = \frac{1}{4}, \text{ and } P(C) = \frac{1}{2}.$$

■

If two events are mutually exclusive, we can compute the probability that one of them occurs from their individual probabilities using a result called the *First Addition Rule*.

Theorem 1.2. *(The First Addition Rule)* If A and B are mutually exclusive events, then

$$P(A \text{ or } B) = P(A) + P(B).$$

Example 1.4. A Pass line bet in craps wins on the first roll if the dice show a sum of either 7 or 11. These are mutually exclusive sums, so we have

$$P(\text{Win on first roll}) = P(7 \text{ or } 11) = P(7) + P(11) = \frac{6}{36} + \frac{2}{36} = \frac{8}{36}.$$

■

If A and B are not mutually exclusive, a slightly more complicated formula can be used to calculate $P(A \text{ or } B)$.

Theorem 1.3. *(The Second Addition Rule)* If A and B are any two events, then

$$P(A \text{ or } B) = P(A) + P(B) - P(A \text{ and } B).$$

Proof. By definition,

$$P(A \text{ or } B) = \frac{\#(A \text{ or } B)}{\#(\mathbf{S})}.$$

What we need to do is compute $\#(A \text{ or } B)$. Elements of A or B can be counted by adding together the number of elements of A and of B, but if they belong to both, they have just been counted twice. In order that each element is only counted once, we must subtract the number of elements that belong to both A and B, that is, the number of elements in the intersection of A and B. This gives

$$\#(A \text{ or } B) = \#(A) + \#(B) - \#(A \text{ and } B).$$

Dividing by $\#(\mathbf{S})$ converts these counts into probabilities, completing the proof.　　　　　　　　　　　　　　　　　　　　　　　　　　　　　　□

The First Addition Rule is a special case of the Second, for if A and B are mutually exclusive, then they cannot occur together, meaning that $P(A \text{ and } B) = 0$.

One obvious pair of mutually exclusive events is an event A and the contrary event that A does not occur, which is called the *complement* of A and denoted A^C.

Example 1.5. If we draw one card from a standard deck and A is the event "An ace is drawn," the complement would be the event A^C: "The card drawn is not an ace."　　　　　　　　　　　　　　　　　　　　　　　　　　■

We then have the following immediate corollary of the First Addition Rule called the *Complement Rule*.

Theorem 1.4. *(The Complement Rule) Let A be an event. Then*

$$P(A^C) = 1 - P(A).$$

Proof. Given any event A, we know from the definition of "complement" that A and A^C have no elements in common, hence A and A^C are mutually exclusive events. The First Addition Rule then states that

$$P(A \text{ or } A^C) = P(A) + P(A^C).$$

Since the two events A and A^C together cover the entire sample space $\mathbf{S}$, *every* outcome lies in exactly one of the two events A and A^C, and so

$$P(A \text{ or } A^C) = P(\mathbf{S}) = \frac{\#(\mathbf{S})}{\#(\mathbf{S})} = 1.$$

Combining these two equations gives $P(A) + P(A^C) = 1$, and rearranging yields $P(A^C) = 1 - P(A)$, as desired.　　　　　　　　　　　　　□

The Complement Rule is sometimes useful in simplifying calculations, where it can reduce a great number of computations to one.

Example 1.6. To compute the probability of turning a profit in 30 spins of an American roulette wheel, making a \$1 wager on the number 14 each time, we need to find the probability of winning at least once in 30 spins. This can be computed by calculating

$$P(1) + P(2) + \cdots + P(30),$$

where $P(x)$ denotes the probability of winning x times in 30 spins. It is far simpler to use the Complement Rule and find

$$1 - P(0),$$

which yields the same value: approximately .5507. ∎

1.2 Conditional Probability

Definition 1.2. Two events A and B are *independent* if the occurrence of one has no effect on the occurrence of the other one.

Example 1.7. On a roll of the dice at a craps table, the two dice fall and bounce on their own, with no consistent influence on each other. The probability of the second die showing a 6 is not affected by whether or not a 6 shows on the first die, so we may regard the numbers on the dice as independent outcomes. ∎

Two events that are mutually exclusive (Section 1.1) are explicitly *not* independent, since the occurrence of one eliminates the chance of the other occurring. Moreover, two events that are independent cannot be mutually exclusive.

It is a fundamental principle of gambling mathematics that *successive trials of random experiments are independent.* This includes successive die rolls at craps, successive wheel spins at roulette, and successive weekly drawings of six Powerball numbers, but *not* successive hands in blackjack—for in blackjack, a card played in one hand is a card that cannot be played in the next hand. Since the composition of the deck has changed, we are not considering successive trials of the same random experiment.

This principle of independence is not always well understood by gamblers, and the inability or unwillingness to understand the doctrine of independent trials is sometimes called the *Gambler's Fallacy.* This fallacy is commonly committed by roulette players who have too strong a belief in the Law of Large Numbers, although it can crop up in any game where successive trials are independent.

Example 1.8. Many casinos have a lighted display near their roulette tables that shows the results of the last 10–20 spins. While some people use these lists as an aid to picking their numbers, in the belief that "hot" numbers, or numbers that appear frequently, are more likely to be drawn again, there is no mathematical advantage to doing so. Assuming that the wheel is not defective in some way, all possible roulette numbers are equally likely; this equiprobability is key to the success of roulette as a casino game. ∎

Balls, dice, wheels, and cards don't understand the laws of probability. They have no knowledge of the mathematics we humans have devised to describe their actions, and they certainly don't understand what the long-term distribution of results is supposed to be. For the same reason, in Example 1.8, it would be equally erroneous to bet on the "cold" numbers that haven't appeared recently, on the grounds that they're somehow "due."

If A and B are independent events, it is a simple matter to compute the probability that they occur together, with the use of a theorem called the *Multiplication Rule*.

Theorem 1.5. *(Multiplication Rule)* *If A and B are independent events, then*

$$P(A \text{ and } B) = P(A) \cdot P(B).$$

Informally, the Multiplication Rule states that we can find the probability that two successive independent events occur by multiplying the probability of the first by the probability of the second.

Example 1.9. Continuing Example 1.7: When rolling two dice, the probability of rolling two 6s is simply the product of the probability of rolling a 6 on the first die and the probability of rolling a 6 on the second die. We have

$$P(12) = P(6) \cdot P(6) = \frac{1}{6} \cdot \frac{1}{6} = \frac{1}{36}.$$

∎

The Multiplication Rule can be extended to any finite number of independent events: the probability of a sequence of n independent events is simply the product of the n probabilities of the individual events.

Example 1.10. In many state lotteries' Daily 4 drawings, the machines used to draw each digit are separate, often four separate clear globes containing 10 ping-pong balls numbered from 0–9 and mixed using a powerful air blower. As a result, the individual digits are independent of one another. A wager on a single 4-digit number, such as 1729, has 1 chance in 10,000 of winning; this can be seen by looking at the probabilities of the four digits:

$$P(\text{Win}) = \left(\frac{1}{10}\right)^4 = \frac{1}{10^4} = \frac{1}{10,000}$$

—just what we calculate by thinking of the number 1729 as one number among 10,000 possibilities. ∎

If the events A and B are not independent, we will need to generalize Theorem 1.5 to handle the new situation. This generalization requires the idea of *conditional probability*. We begin with some examples.

Example 1.11. If we draw one card from a standard deck, the probability that it is a king is $\frac{4}{52} = \frac{1}{13}$. If, however, we are told that the card is a face card, the probability that it's a king is $\frac{4}{12} = \frac{1}{3}$—that is, additional information has changed the probability of our event by allowing us to restrict the sample space. If we denote the events "The card is a king" by K and "The card is a face card" by F, this last result is written $P(K|F) = \frac{1}{3}$ and read as "the (conditional) probability of K given F is $\frac{1}{3}$." ∎

Example 1.12. In Example 1.10, the probability of 1729 being the winning number in a Daily 4 lottery drawing was found to be 1/10,000. If the first digit is drawn and found to be 1, the probability that 1729 will win has risen to 1/1000. Of course, if the first digit is drawn and is not 1, the probability of 1729 winning has dropped all the way to 0. ∎

The fundamental idea here is that more information can change probabilities. If we know that the event A has occurred and we're interested in the event B, we are now looking not for $P(B)$, but $P(B \text{ and } A)$, because only the part of B that overlaps with A is possible. With that in mind, we have the following formula for conditional probability:

Definition 1.3. The *conditional probability* of B given A is

$$P(B \,|\, A) = \frac{P(B \text{ and } A)}{P(A)}.$$

This formula divides the probability of the intersection of the two events by the probability of the event that we know has already occurred. Note that if A and B are independent, we immediately have $P(B\,|\,A) = P(B)$, since then $P(B \text{ and } A) = P(A) \cdot P(B)$. This is one case where more information—in this case, the knowledge that A has occurred—does not change the probability of B occurring.

Example 1.13. When betting on Red in two successive spins of a roulette wheel, the outcome on the first spin has no effect on the result of the second spin. Losing or winning on the first spin neither raises nor lowers the chance of winning on the second spin, since successive trials are independent. ∎

As with the addition rules, we can state a second, more general, version of the Multiplication Rule that applies to any two events—independent or not—and reduces to the first rule when the events are independent. This more general rule simply incorporates the conditional probability of B given A, since we are looking for the probability that both occur.

Theorem 1.6. *(General Multiplication Rule) For any two events A and B, we have*
$$P(A \text{ and } B) = P(A) \cdot P(B \,|\, A).$$

Proof. This result follows from the fact that $P(A \text{ and } B) = P(B \text{ and } A)$ and from Definition 1.3. □

Table 1.1 collects some of the more useful probability rules that we have discussed so far. In referring to this table, note that the First Addition Rule applies only to mutually exclusive events, while the Second Addition Rule applies to all events. Similarly, the first Multiplication Rule in the table requires that the events A and B be independent; the General Multiplication Rule applies in all cases.

TABLE 1.1: Probability Rules

First Addition Rule **(mutually exclusive events)**	$P(A \text{ or } B) = P(A) + P(B).$
Second Addition Rule	$P(A \text{ or } B) = P(A) + P(B) - P(A \text{ and } B).$
Complement Rule	$P(A^C) = 1 - P(A).$
Multiplication Rule **(independent events)**	$P(A \text{ and } B) = P(A) \cdot P(B).$
Conditional Probability	$P(B \mid A) = \dfrac{P(A \text{ and } B)}{P(A)}.$
General Multiplication Rule	$P(A \text{ and } B) = P(A) \cdot P(B \mid A).$

1.3 Combinatorics

Combinatorics is the branch of mathematics that studies counting techniques. In many applications of the definition of probability given on page 1, the sheer number of elements comprising an event or the sample space is far too large to count them one by one. When computing probabilities, we seldom have need to consider each of these simple events individually; we are usually only interested in how many there are. Frequently in gambling mathematics, we find ourselves considering the number of ways in which several events can happen in sequence. If we know the number of ways that each individual event can happen, elementary combinatorics tells us that simple multiplication can be used to find the answer.

Fundamental Counting Principle

Theorem 1.7. *(Fundamental Counting Principle) If there are n independent tasks to be performed, such that task T_1 can be performed in m_1 ways, task T_2 can be performed in m_2 ways, and so on, then the number of ways in which all n tasks can be performed successively is*

$$N = m_1 \cdot m_2 \cdot \ldots \cdot m_n.$$

That the Fundamental Counting Principle (FCP for short) is a reasonable result can be easily seen by testing out some examples with small numbers and listing all possibilities—for example, when rolling 2 six-sided dice (abbreviated 2d6), one red and one green, each die is independent of the other and can land in any of six ways. By the FCP, there are $6 \cdot 6 = 36$ ways for the two dice to fall, and this may be confirmed by writing out all of the possibilities.

Example 1.14. If an experiment consists of drawing one card from a standard deck and then rolling a d6, there are $52 \cdot 6 = 312$ different possible outcomes. ∎

A special case of the Fundamental Counting Principle arises when we consider the number of ways to arrange a set of n elements, with no repetition allowed, in different orders. The first element may be chosen in n ways, the second in $n - 1$, and so on, down to the last item, which may be chosen in only one way. The total number of orders for a set of n elements is thus $N = n \cdot (n - 1) \cdot (n - 2) \cdot \ldots \cdot 3 \cdot 2 \cdot 1$. This number is called n *factorial* and written $n!$.

Definition 1.4. If n is a natural number, the *factorial* of n, denoted $n!$, is the product of all of the positive integers up to and including n:

$$n! = 1 \cdot 2 \cdot 3 \cdot \ldots \cdot (n - 1) \cdot n.$$

$0! = 1$, by definition.

It is an immediate consequence of the definition that $n! = n \cdot (n - 1)!$. Factorials get very big very fast. $4! = 24$, but then $5! = 120$ and $6! = 720$.

Example 1.15. A standard deck of cards may be permuted in $52!$ different ways. This number is approximately 8.066×10^{67}. ∎

Example 1.16. In the 1970s and 1980s, the largest factorial that could be computed on a standard scientific calculator whose number capacity was limited to numbers less than 10^{100} was

$$69! \approx 1.7112245 \times 10^{98},$$

a limit whose first 7 decimal places were well known to many people whose interest in mathematics was generated in part by playing with early LED calculators. More modern calculators such as the TI-nSpire that can handle numbers up to 10^{1000} can compute factorials through

$$449! \approx 3.8519305 \times 10^{997}.$$

∎

Where Order Matters: Permutations

Definition 1.5. A *permutation* of r items from a set of n items is a selection of r items chosen so that the order matters.

For example, ABC is a different permutation of three alphabet letters than CBA. It should be noted that "order" may appear in several forms. One way to determine whether or not order matters in making a selection is to ask if different elements of the selection are being treated differently once they are chosen.

Example 1.17. The *trifecta* wager at horse racetracks calls for bettors to select the first 3 horses to finish the race, in order. If 8 horses start the race, the FCP can be used to show that there are $8 \cdot 7 \cdot 6 = 336$ different possible trifecta tickets. ∎

Most casino games do not require that events occur in a specified order. One exception was the Rocket Bet at *Rocket Roulette*, a game variation played with a modified American wheel for a brief while in the 1970s at the El Cortez Casino in Las Vegas. The wheel replaced numbers with letters and symbols; it was designed by Lucille Farrlow and named the "American Wheel of Fortune" [5]. The Rocket Roulette layout is shown in Figure 1.1.

Rocket Roulette appeared just before America's bicentennial, and so the symbols and color scheme had a patriotic flavor. Red and black spaces were replaced by red, white, and blue; each color was assigned to 12 spaces. The zeros were colored gold and displayed a map of the continental United States and a bald eagle.

Most Rocket Roulette wagers were direct translations of standard American roulette bets to the new wheel. Betting options 1–6 in Figure 1.1 are simple bets on 1–6 adjacent symbols, and pay off at the same rate as on an American wheel: if n symbols are selected, the payoff is

$$\frac{36 - n}{n} \text{ to } 1$$

unless $n = 5$. Bets on 5 symbols pay 6–1.

Option 7 extends the 5-number basket bet to other nonadjacent groups of 5 symbols: the solid symbols (bell, airplane, elephant, heart, and telephone) or the outline symbols (two-fingered peace sign, donkey, car, peace symbol, and American flag). The dozens bets on a Rocket wheel (option 8) cover columns, colors, groups of solid or outline letters, or blocks of 12 consecutive symbols, dubbed Atlantic, Gulf and Pacific instead of Low, Middle, and High. Since there are no odd or even symbols and the colors on a Rocket Roulette wheel are in groups of 12 rather than 18, even-money bets (option 9) are restricted to the first and last 18 symbols on the layout; these are the Rebel and Yankee bets.

Betting option #10, the Rocket Bet, was a new wager where the order mattered. This was a wager that the eagle, map, and American flag would

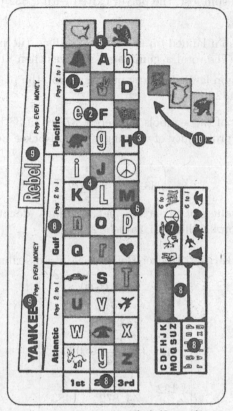

© Lucille Farrlow—1973 "American Wheel of Fortune"™

FIGURE 1.1: Rocket Roulette layout and betting options. Image reprinted from *Systems & Methods* [5]; used with permission.

turn up *in that order* on three consecutive spins. (The analog on an American roulette wheel would be three straight spins landing on 00, 0, and 9. This has far less patriotic significance.) This bet progressed up the rocket as each spin won.

- A $1 Rocket bet paid off at 35–1, the standard one-number payoff, if the first spin was the bald eagle. $33 of the $36 total was collected by the gambler and the remaining $3 advanced to the next wager. The player's profit at this point is $32.

- If the second spin was the map of the continental USA, the payoff was 40 for 1 (39–1 plus the return of the $3 bet), for a total of $120. $100 was collected as profit (bringing the total profit to $132) and the remaining $20 moved on to the final bet.

If the second spin lost, the player exited with a $32 profit from the first spin.

- If the third spin landed on the American flag, the payoff was 50 for 1: $1000 in all. The total accumulated win was then $1132.

If the third spin landed on any other symbol, the player could still walk away with $132 in profit.

Since the order of the symbols mattered, 37 out of 38 Rocket Bets lost on the first spin. The probability of winning the top Rocket Bet prize was simply

$$\left(\frac{1}{38}\right)^3 = \frac{1}{54,872}.$$

The possible values of a player's net winnings, denoted here by x, are shown with their probabilities in Table 1.2.

TABLE 1.2: Rocket Bet probability distribution

x	$P(x)$
-1	$\dfrac{37}{38} \approx .9737$
32	$\dfrac{37}{1444} \approx .0256$
132	$\dfrac{37}{54,872} \approx \dfrac{1}{1483}$
1332	$\dfrac{1}{54,872}$

We are usually not as interested in a list of all of the permutations of a set as in how many permutations there are. The following theorem allows easy calculation of that number, which is denoted $_nP_r$.

Theorem 1.8. *The number of permutations of r items chosen from a set of n items is*

$$_nP_r = \frac{n!}{(n-r)!}.$$

Proof. There are n ways to select the first item. Once an item is chosen, it cannot be chosen again, so the second item may be chosen in $n-1$ ways. There are then $n-2$ items remaining for the third choice, and so on until there are $n-r+1$ numbers remaining from which to choose the rth and final term. By the Fundamental Counting Principle, we have

$$_nP_r = n \cdot (n-1) \cdot \ldots \cdot (n-r+1).$$

Multiplying the right-hand expression by $1 = (n-r)!/(n-r)!$ gives

$$_nP_r = n \cdot (n-1) \cdot \ldots \cdot (n-r+1) \cdot \frac{(n-r)!}{(n-r)!}$$

$$= \frac{n \cdot \ldots \cdot (n-r+1) \cdot (n-r) \cdot \ldots \cdot 3 \cdot 2 \cdot 1}{(n-r)!}$$

$$= \frac{n!}{(n-r)!}.$$

$\square$

Example 1.18. Rouleno (page 72) is a game where the order of the game elements, in this case 6 pool balls, matters. The number of ways to arrange all of 6 distinguishable items in order is $_6P_6 = 720$. ∎

Where Order Doesn't Matter: Combinations

Most of the time when gambling, we are not so concerned about the order of events, as when a hand of cards is dealt or a set of Powerball numbers is drawn. For counting these arrangements, we are interested in *combinations* rather than permutations.

Definition 1.6. A *combination* of r items from a set of n items is a subset of r items chosen without regard to order. The number of such combinations is denoted $\binom{n}{r}$, which is read as "n choose r." An alternate notation, in line with the expression counting permutations, is $_nC_r$.

Here, ABC and CBA are indistinguishable combinations, as they are subsets of the alphabet consisting of the same three letters. The different order is not a concern here. If the elements of a selected subset are receiving the same treatment once selected, then the choice is a combination, not a permutation.

Theorem 1.9. *The number of combinations of r items chosen from a set of n items is*

$$\binom{n}{r} = \frac{n!}{(n-r)! \cdot r!} = \frac{_nP_r}{r!}.$$

Proof. We begin with the formula for the number of permutations:

$$_nP_r = \frac{n!}{(n-r)!}.$$

Since we are looking for combinations, two permutations that differ only in the order of the elements are identical to us. Any combination of r elements from a set of n can be rearranged into $r!$ different orders, by the Fundamental Counting Principle. We then have

$$\binom{n}{r} = \frac{_nP_r}{r!} = \frac{n!}{(n-r)! \cdot r!},$$

as desired.

$\square$

Example 1.19. A player wishing to make 7 single-number bets on an American roulette wheel may do so in $\binom{38}{7} = 12,620,256$ ways, as the order in which the numbers are selected and the bets placed does not affect the outcome. ∎

The following theorem collects several simple facts about combinations.

Theorem 1.10. *For all* $n \geqslant 0$:

1. $\binom{n}{0} = \binom{n}{n} = 1$ *and* $\binom{n}{1} = n$.

2. *For all* k, $0 \leqslant k \leqslant n$, $\binom{n}{k} = \binom{n}{n-k}$.

3. $\sum_{r=0}^{n} \binom{n}{r} = 2^n$.

Proof.   1. Given a set of n elements, there is only one way to select none of them—that is, there is only one way to do nothing, so $\binom{n}{0} = 1$. Similarly, since the order does not matter, there is only one way to choose all of the items: $\binom{n}{n} = 1$.

If we are choosing only one item, we may select any element from among the n, and there are thus n choices possible.

2. We note that every selection of k items from a set of n partitions the set into two disjoint subsets: one of size k and the other of size $n - k$, and so choosing k items to take is equivalent to choosing $n - k$ items to leave behind. The conclusion follows immediately.

Alternately, direct application of the formula for combinations gives the following:

$$\binom{n}{k} = \frac{n!}{(n-k)! \cdot k!} = \frac{n!}{k! \cdot (n-k)!} = \frac{n!}{[n-(n-k)]! \cdot (n-k)!} = \binom{n}{n-k}.$$

3. If we think of a combination of r items from a set A with $\#(A) = n$ as choosing a subset of A with r elements, then the left side of this equation is simply the total number of subsets of A of all sizes.

The number of subsets of a set with n elements is easily shown to be 2^n [6]. Since we have counted the set of all subsets of A in two different ways, those two expressions must be the same, completing the proof.

□

Since the formulas for $_nP_r$ and $_nC_r$ involve factorials, which are known to grow very fast, it follows that these numbers also get very big very fast. Starting with a standard deck of 52 cards, the number of possible subsets of a given size r is quite large, even if order is not considered.

Example 1.20. Contract bridge is a four-player game where each player is dealt a hand of 13 cards. The number of possible bridge hands is

$$\binom{52}{13} = 635,013,559,600,$$

more than 600 billion. ■

1.4 Mathematical Expectation

Random Variables

Definition 1.7. A *random variable* (RV for short) is an unknown quantity X whose value is determined by a chance process.

A sequence of examples will illustrate this important idea far better than a formal definition.

Example 1.21. a. Let X be the sum when 2d6 are rolled. X can be any integer value from 2 through 12.

b. Let X be the total accumulated winnings, positive or negative, after 10 consecutive \$1 roulette bets on Red. X is an integer in the range $[-10, 10]$.

c. Let X be the number of aces in a 5-card draw poker hand. $X \in \{0, 1, 2, 3, 4\}$.

d. Let X count the number of matched numbers on a state lottery ticket where players pick, and the state chooses, 6 numbers from 1–47. Then $X \in \{0, 1, 2, 3, 4, 5, 6\}$. ■

Definition 1.8. A *probability distribution* for a random variable X is a list of the possible values of X, together with their associated probabilities.

Probability distributions are most commonly presented in a table of values or as an algebraic formula, which is called a *probability distribution function* or *PDF*.

Example 1.22. If X denotes the number that appears when a fair d6 is rolled, then the probability distribution for X is shown in Table 1.3.

Since the probabilities in the table are all the same, the simple events listed are equally likely. The PDF is simply $P(X = x) = \frac{1}{6}$, which does not depend on x. ■

Example 1.23. If X counts the number of matches on a state lottery ticket where players pick 6 numbers from 1–47 and win a prize if 3 or more of their

TABLE 1.3: Probability distribution when a d6 is rolled

x	$P(X = x)$
1	1/6
2	1/6
3	1/6
4	1/6
5	1/6
6	1/6

numbers are drawn among the state's 6 numbers, then the PDF for X, where $0 \leqslant x \leqslant 6$, can be given algebraically as

$$P(X = x) = \frac{\binom{6}{x} \cdot \binom{41}{6-x}}{\binom{47}{6}}.$$

■

Example 1.24. A bet on a single number at American roulette pays 35–1 if the chosen number is spun. The probability of winning is $\frac{1}{38}$, since there are 38 pockets on the wheel. A random variable X measuring the outcome of a \$1 bet can take on the two values 35 and –1, and Table 1.4 gives the probability distribution for X. ■

TABLE 1.4: Probability distribution for a \$1 single-number American roulette bet

x	$P(X = x)$
35	$\dfrac{1}{38}$
–1	$\dfrac{37}{38}$

Binomial Random Variables

Definition 1.9. A *binomial* experiment has the following four characteristics:

1. The experiment consists of a fixed number of successive trials, denoted by n.

2. Each trial has exactly two outcomes, denoted *success* and *failure*.

   Practically speaking, it is often possible to collect multiple outcomes together in order to reduce the number of outcomes down to 2. For example, in considering a 4-number wager at European roulette, we can denote the 4 selected numbers as "success" and the other 33 numbers as "failure"—if we lose our bet, the number that was actually spun doesn't really matter.

3. The probabilities of success and failure are constant from trial to trial. We denote the probability of success by p and the probability of failure by q, where $q = 1 - p$.

4. The trials are independent.

A random variable X that counts the number of successes of a binomial experiment is called a *binomial* random variable. The values n and p are called the *parameters* of X.

Example 1.25. Consider a sequence of 10 consecutive bets on Red at American roulette. The number of wins X is a binomial random variable with parameters $n = 10$ and $p = \frac{18}{38}$. ∎

Suppose that X is a binomial random variable with parameters n and p. We can derive a PDF for X by thinking of the n trials as a sequence of n boxes and designating each one as either "success" or "failure." To find $P(X = k)$:

- Begin by selecting which k of the n boxes represent the successes. This may be done in $\binom{n}{k}$ ways.

- The remaining $n - k$ boxes must then be filled in with failures.

- Since the trials are independent, the probability of k successes is p^k and the probability of $n - k$ associated failures is $(1 - p)^{n-k} = q^{n-k}$.

- Independence allows us to conclude that the probability of a specific sequence of k successes and $n - k$ failures is then $p^k \cdot q^{n-k}$.

- Multiplication by the number of ways to choose such a sequence gives

$$P(X = k) = \binom{n}{k} \cdot p^k \cdot q^{n-k}.$$

Expected Value and House Advantage

The notion of *expected value* is fundamental to any discussion of random variables and is especially important when those random variables arise from a game of chance. The expected value of a random variable X is, in some sense, an average value, or what we might expect in the long run if we were to sample many values of X.

The common notion of "average" corresponds to what statisticians call the *mean* of a set of numbers: add up all of the numbers and divide the sum by how many numbers there are. For a random variable X, this approach requires some fine-tuning, as there is no guarantee that a small sample of values of X will be representative of the range of possible values. Our interpretation of average will incorporate each possible value of X together with its probability.

Definition 1.10. Let X be a random variable with a given probability distribution function $P(X = x)$. The *expected value* or *expectation* $E(X)$ of X is computed by multiplying each possible value for X by its corresponding probability and then adding the resulting products:

$$E(X) = \sum_x x \cdot P(X = x).$$

This expression may be interpreted as a standard mathematical mean drawn from an infinitely large random sample. If we were to draw such a sample, we would expect that the *proportion* of sample elements with the value x would be $P(X = x)$; adding up over all values of x gives this formula for $E(X)$.

We may abbreviate $E(X)$ as E when the random variable is clearly understood. The notation $\mu = E(X)$, where μ is the Greek letter mu, is also common, particularly when the expected value appears as a term in another expression.

The notation used in Definition 1.10 does not indicate the limits of the indexing variable x, as is customary with sums; this is because those values may not be a simple list running from 1 to some n. When written this way, we should take the sum over *all* possible values of the random variable X, as in the next example.

Example 1.26. Let X denote the net win when a \$1 single-number American roulette wager is placed, as we saw in Example 1.24. The expected value of X is then

$$E = (35) \cdot \frac{1}{38} + (-1) \cdot \frac{37}{38} = -\frac{1}{19} \approx -\$.0526,$$

or about −5.26¢.   ∎

Since this expected value is negative, the player can expect to lose money on an average spin, to the tune of 5.26¢ per dollar wagered. The rules of every table game have been set to ensure a negative player expectation, and thus a game that favors the casino.

If a random variable is binomial, computing its expected value is simple.

Theorem 1.11. *If X is a binomial random variable with parameters n and p, then $E(X) = np$.*

The expected number of successes is just the number of trials multiplied by the probability of success on a single trial.

Proof.

$$E(X) = \sum_{x=0}^{n} x \cdot P(X = x)$$

$$= \sum_{x=0}^{n} x \cdot \binom{n}{x} \cdot p^x \cdot q^{n-x}$$

$$= \sum_{x=0}^{n} x \cdot \frac{n!}{(n-x)! \cdot x!} \cdot p^x \cdot q^{n-x}.$$

Since the $x = 0$ term is equal to zero, we can drop that term from the sum and renumber starting at 1:

$$E(X) = \sum_{x=1}^{n} x \cdot \frac{n!}{(n-x)! \cdot x!} \cdot p^x \cdot q^{n-x}$$

$$= \sum_{x=1}^{n} \frac{n!}{(n-x)! \cdot (x-1)!} \cdot p^x \cdot q^{n-x}$$

$$= np \cdot \sum_{x=1}^{n} \frac{(n-1)!}{(n-x)! \cdot (x-1)!} \cdot p^{x-1} \cdot q^{n-x}$$

$$= np \cdot \sum_{x=1}^{n} \frac{(n-1)!}{[(n-1)-(x-1)]! \cdot (x-1)!} \cdot p^{x-1} \cdot q^{n-1-x+1}.$$

If we substitute $y = x - 1$ in this last sum, we have

$$E(X) = np \cdot \sum_{y=0}^{n-1} \frac{(n-1)!}{[(n-1)-y]! \cdot y!} \cdot p^y \cdot q^{n-1-y}$$

$$= np \cdot \sum_{y=0}^{n-1} P(Y = y).$$

This sum is the sum of all of the probabilities of a binomial random variable Y with parameters $n-1$ and p, and so adds up to 1, completing the proof. □

A concept related to expected value is the *house advantage* associated with a game of chance.

Definition 1.11. The *house advantage (HA)* of a game with a wager of N and payoffs given by the random variable X is $-E(X)/N$.

If the expectation is negative, as it is in virtually every casino game, the HA will be positive. The house advantage of a game is frequently expressed as a percentage of the original wager, so the \$1 roulette wager above can be said to have a house advantage of 5.26%. This value is commonly used by game designers and players as a benchmark dividing good and bad bets: a wager with a HA exceeding 5.26% may give the casino too much of an edge to be viable unless other factors, such as a high potential payoff for a small wager, are present. The HA measures how much of the total amount wagered can be reliably expected to be retained by the casino or lottery agent, in the long run.

Example 1.27. Using the probability distribution in Table 1.2, one can show that the house advantage on Rocket Roulette's Rocket Bet is a low-for-roulette 4.05%. This lower house edge is achieved by paying off the second and third bets at better than 35–1.   ∎

Definition 1.12. If X is a random variable measuring the payoffs from a game, we say that the game is *fair* if $E(X) = 0$.

Equivalently, a fair game has zero house advantage.

If a game is fair, then in the long run, we expect to win exactly as much money as we lose, and thus, aside from any possible entertainment derived from playing, we expect no gain. A sensible approach to fair and unfair games is often summarized in the following maxim:

> If a game is fair, don't bother to play.
> If a game is unfair, make sure it's unfair in your favor.

If all gamblers followed this sound principle, casinos and lotteries would cease to exist. Player failure to follow this advice is responsible for the ongoing success of the gambling industry, for games which are unfair and favor the gambler are rare. Usually, such games are the result of faulty design, a mathematical error in the evaluation process, or an exploitable rule that can be attacked by *advantage players*: gamblers who look for vulnerable casino games that can be beaten within the rules. This was the case, for example, with Pell (page 143).

As an informal guide to gambling, it is often said that "the easier a bet is to understand, the higher the house advantage."

Example 1.28. The Big Six wheel (page 62) has a simple rule: pick a symbol, and you win if that symbol turns up when the wheel is spun. Every bet on the wheel carries a high HA: from 11–24%.

Baccarat, on the other hand, has complicated rules for drawing and standing, and the HAs on the Banker and Player bets are both under 1.25%.   ∎

An Extended Example: Red Dragon

Red Dragon is a game played with 15d6 that may have had an origin in actual carnivals as well as in casinos [94]. Players bet on a number from 1–6, and then the 15 dice are rolled. A gambler wins if his or her selected number appears on 3 or more of the dice. If 3–5 dice show the player's number, the payoff is 1–1; if 6 or more dice show the number, the winning payoff is 2–1.

For convenience, suppose that the player has chosen the number 4. The number of 4s on the 15 dice is a binomial random variable X with parameters $n = 15$ and $p = \frac{1}{6}$, so the expected number of 4s on a single toss is $\frac{15}{6} = 2.5$. This expected value applies, of course, to any number from 1–6. Winning a Red Dragon bet requires a better-than-average roll. Since it is impossible to roll any number exactly 2½ times, it is easily seen that there must be at least one winning number on every throw of the dice. This fact could be useful in promoting the game.

The probability of winning \$1 is simply $P(X = 3) + P(X = 4) + P(X = 5)$, or

$$\sum_{k=3}^{5} \binom{15}{k} \cdot \left(\frac{1}{6}\right)^{k} \cdot \left(\frac{5}{6}\right)^{15-k} \approx .4404.$$

Similarly, the probability of winning \$2 is

$$\sum_{k=6}^{15} \binom{15}{k} \cdot \left(\frac{1}{6}\right)^{k} \cdot \left(\frac{5}{6}\right)^{15-k} \approx .0274.$$

By the Complement Rule, the probability of losing a Red Dragon bet is approximately .5322—as we expected, slightly more than ½.

The HA of a Red Dragon bet is then 3.71%.

Red Dragon allows for multiple winning numbers on a single throw of the dice, as when five 3s, seven 4s, and three 6s are rolled and three numbers win. Suppose, for example, that we wish to find the probability that 5 of the 6 numbers win on a single throw. For this to happen, 5 numbers need to be rolled 3 times each. Once the 5 numbers have been selected, we need to count the number of ways to arrange 3 copies of each number among 15 spaces. This question asks for the number of *distinguishable permutations* of 15 symbols. Permutations that differ from one another only by rearranging copies of the same number are not distinguishable. It may be useful to think of the 3 copies of each number as appearing in different colors: red, green, and blue. Rearranging the red, green, and blue 4s does not produce a mathematically different permutation of the 15 digits involved.

Theorem 1.12. *Suppose that there are N items to be permuted, of which n_1 are of type 1, n_2 of type 2, and so on up to n_m objects of type m, where $n_1 + n_2 + \cdots + n_m = N$. Then the number of distinguishable permutations of the n items is*

$$\frac{N!}{n_1! \cdot n_2! \cdots n_m!}.$$

Proof. Despite the use of the word "permutations," which suggests that order matters, deriving this formula is more a matter of selecting the places where items fall without regard to order. There are $\binom{N}{n-1}$ ways to select the positions for the n_1 elements of type 1. Once this has been done, there are $N - n_1$ open positions and n_2 objects of type 2 to be placed. Accordingly, there are $\binom{N-n_1}{n_2}$ ways to set the n_2 type-2 elements. At this point, the Fundamental Counting Principle shows that there are

$$\binom{N}{n_1} \cdot \binom{N-n_1}{n_2} = \frac{N!}{(N-n_1)! \cdot n_1!} \cdot \frac{(N-n_1)!}{(N-n_1-n_2)! \cdot n_2!}$$

$$= \frac{N!}{n_1! \cdot n_2! \cdot (N-n_1-n_2)!}$$

ways to place the first $n_1 + n_2$ items.

Continuing, there are

$$\frac{(N-n_1-n_2)!}{(N-n_1-n_2-n_3)! \cdot n_3!}$$

ways to choose the positions of the type-3 items, and so forth. At the end of this selection process, we seek to place n_{m-1} elements of type $m-1$ into $N - n_1 - n_2 - \cdots - n_{m-2}$ slots, so we have a term of the form

$$\frac{(N-n_1-n_2-\cdots-n_{m-2})!}{(N-n_1-n_2-\cdots-n_{m-2}-n_{m-1})! \cdot n_{m-1}!}.$$

After this, there remain n_m type-m items to be placed in

$$N - n_1 - n_2 - \cdots - n_{m-2} - n_{m-1}$$

positions, and that can be done in only one way. Multiplying each of these terms together gives

$$\binom{N}{n_1} \cdot \binom{N-n_1}{n_2} \cdot \ldots \cdot \binom{N-n_1-n_2-\cdots-n_{m-2}}{n_{m-1}} = \frac{N!}{n_1! \cdot n_2! \cdot \ldots \cdot n_m!},$$

the desired result.   □

For this question, we have

$$\frac{15!}{(3!)^5} = 168,168,000$$

distinguishable permutations of 3 copies of each of the 5 winning numbers. There are $\binom{6}{5} = 6$ ways to pick 5 winning numbers, and so we have

$$P(5 \text{ winning numbers}) = \binom{6}{5} \cdot \frac{15!}{(3!)^5} \cdot \left(\frac{1}{6}\right)^{15} \approx .0021 \approx \frac{1}{466}.$$

Example 1.29. In how many ways can a single roll of the 15 dice produce 2 numbers that pay off at 2–1?

There are 8 types of distinguishable permutations to count here.

n_1	n_2	n_3	n_4	n_5
6	6	1	1	1
6	6	2	1	
6	6	3		
7	6	1	1	
7	6	2		
7	7	1		
8	6	1		
8	7			

For each type, the number of distinguishable permutations is

$$\frac{15!}{n_1! \cdot n_2! \cdot n_3! \cdot n_4! \cdot n_5!},$$

where $n_k = 0$ and $n_k! = 1$ if a slot in the chart above is vacant. This number must then be multiplied by the number of ways to assign numbers from 1–6 to each of the outcomes on a single roll. For the 6-6-2-1 distribution, this number is $\binom{6}{2} \cdot 4 \cdot 3 = 180$, and so there are

$$\frac{15!}{6! \cdot 6! \cdot 2! \cdot 1!} \cdot 180 = 1,261,260 \cdot 180 = 227,026,800$$

ways to roll 15 dice and get a 6-6-2-1 distribution. Repeating this calculation for the other 7 distributions gives a total of 498,776,850 ways to roll 15 dice and have 2 numbers appearing 6 or more times each.

The probability of two 2–1 payoffs is then

$$\frac{498,776,850}{6^{15}} \approx \frac{1}{943}.$$

■

Table 1.5 shows the house advantages for many of the games considered in this book.

TABLE 1.5: House advantages for some casino carnival games

Game	HA (%)	Page
Rouleno, single-number bet	1.67	72
Triple Shot, Blackjack bet	1.84	135
Spider Craps, Don't Pass Line	2.58	30
Triple Shot, War bet	2.71	135
Dice Duel, Red and Gold bets	2.78	225
Triple Shot, Poker bet	3.20	135
Die Rich	3.70	198
Red Dragon	3.71	21
Mystery Card Roulette, Suit bet	3.70	66
Rocket Roulette, Rocket bet	4.05	10
Street Dice, main bet	5.02	212
Street Dice, Brick Bet	6.27	215
Pell, suit bets	7.68	144
Chuck-a-luck	7.87	229
Casino Merry-Go-Round, bet A	9.00	38
Beat the Dealer	11.11	206
La Boule, all bets	11.11	59
Pokette, 2-card Pair bet	11.11	68
Sic bo, Straight bet	11.11	232
Let It Ride (1-die game)	12.50	201
Faro, calling the turn	16.67	159
Big Six Wheel, wager on $10 spot	18.52	62
High Roll Dice, "Best longshot bet"	74.20	276
Repeater craps bet on 12, Table 2.2	99.84	28

Chapter 2

Mathematics and Casino Game Design

2.1 Introduction

Before a carnival game can reach the casino floor, it must be approved by gaming officials. The approval process includes a report from a gaming mathematician attesting to the odds of winning and other important probabilities that are involved in the game, which is typically reviewed for accuracy by outside evaluators. In this chapter, we shall examine some of the calculations that these mathematicians are called on to make and see how these numbers allow room for game designers to adjust parameters to improve an initial game design.

In addition to relatively simple calculations like the probabilities of a player win and house win, these computations can include the house edge on main wagers and on side bets, the effects on the HA of minor or major rule changes, and analysis of the effects that advantage play such as card-counting might have on the game [50].

2.2 House Advantage

It is certainly true that careful calculation of the house advantage of every available wager is an important stage in the development of a new casino game. Some gaming experts suggest the following ranges of HAs as worthy targets for game designers [45, 50]:

- For a main wager: Anywhere from .5–6%. American roulette, with its 5.26% HA on every wager but one, is a useful benchmark here. If a game has a higher HA than American roulette on its primary wager, it is unlikely to attract enough players to succeed.

- For a *side bet*, an optional wager available for a game that is peripheral to the play of the main game: 2–9%. As optional wagers, side bets can be avoided by gamblers who understand or can quickly evaluate the mathematics involved.

- For a wager linked to a *progressive jackpot*, a variable prize that grows by adding a small percentage of each losing wager until the jackpot is won: as much as 25%. These wagers couple a low probability of winning with a large—often life-changing—jackpot.

A game whose primary bet is worse for the player than American roulette may be viable if the game in question has other compelling features. One of these may be the possibility of a large jackpot for a very small investment; this keeps keno in the mix at many casinos. While the house edge for keno typically runs at 25% or higher, a gambler isn't going to turn $1 into $25,000 or more on a single wager playing blackjack. This combination of a high house edge with a potential high payoff is also applicable to many casino side bets. For example, some blackjack tables offer a "Royal Match" bet that pays off if the player's first two cards are suited, with a bonus if the cards are a suited king and queen [6]. This bet, which carries a 10.9% HA, has nothing to do with blackjack's main objective: to get closer to 21 than the dealer without going over.

Another possible compelling feature is an easy-to-play game. The Big Six Wheel (page 62) is easy to play and to understand, and at many casinos, it sits near the edge of the gaming floor to beckon potential gamblers with a game that has the promise of being entertaining, but not overwhelming. These friendly features mask the game's 11–24% HA. This edge is a price that gamblers pay for simplicity; a casino patron willing to learn blackjack basic strategy or approach the often-intimidating baccarat table is rewarded with far lower HAs, on the order of 1% or less.

Setting a suitable HA involves careful choices among the variables that are part of any game's design. We consider three places in the design process where choices affect the HA.

1. The parameters of the game, including the rules and the equipment used.

Some new craps bets act, in part, to maintain interest in a craps game when the shooter holds the dice through many rolls without making a point or sevening out. On that list, one may find the *Repeater* bet, a new wager developed and marketed by Aces Up Gaming, and offered at several casinos in Las Vegas and elsewhere. To make the simplest Repeater bet, the gambler selects a number from 2–12, excluding 7, and is paid according to Table 2.1 if that number is thrown a specified target number of times before a 7 is rolled. If n is the number chosen by the player, the target number is n if n is from 2–6 and $14 - n$ if n is between 8 and 12. Craps dealers are supplied with markers to be placed on the modified layout to track how many times each number is rolled.

TABLE 2.1: Repeater bet target numbers and payoff odds

Roll	Target number	Payoff odds
2, 12	2	40 for 1
3, 11	3	50 for 1
4, 10	4	65 for 1
5, 9	5	80 for 1
6, 8	6	90 for 1

Note that the payoffs take the form X *for* 1 rather than X *to* 1; this means that the payoff includes the return of the player's original wager. This is common practice on craps tables; it makes wagers look slightly more lucrative than they really are. A payoff of X for 1 corresponds to payoff odds of $(X-1)$ to 1.

An additional Repeater bet option covers the Any Craps numbers plus 11: 2, 3, 11, and 12. If these numbers, in any combination or order, are rolled a total of 7 times between 7s, the payoff is 100 for 1. The payoffs on Repeater bets are higher than the standard craps bets, whose payoffs top out at 30–1, offer. These high payoffs are designed to draw in new players to the craps table by offering the potential for big wins on a par with slot machines, where casual gamblers might be found and introduced to dice [14].

A Repeater bet may be placed prior to a come-out roll or at any time during a string of rolls provided that the selected number has not yet been rolled. On a 7, all Repeater bets lose unless another come-out roll has been reached; all Repeater bets are "off," or temporarily inactive, during a come-out roll. Rolling an 8 on a come-out roll does not count as a roll of 8 toward the goal of six 8s, but neither does a come-out roll of 7 end the bet with a loss.

For a given Repeater bet, let p be the probability of a successful roll, and let $q = \frac{5}{6} - p$ be the probability of a "neutral" roll that is neither a selected number nor 7. The probability of winning a Repeater bet is the probability that the chosen number is rolled the necessary number of times, k, before a 7 is rolled. Since the number of rolls is theoretically unlimited, the probability is given by an infinite series:

$$P(\text{Win}) = \sum_{i=k}^{\infty} P(\text{Win in } i \text{ rolls})$$

$$= \sum_{i=k}^{\infty} \binom{i-1}{k-1} \cdot p^k q^{i-k}.$$

The factor $\binom{i-1}{k-1}$ reflects that fact that $k-1$ of the first $i-1$ rolls must be chosen as winners, leaving the last roll to be the selected number and thus guaranteeing a win in exactly i rolls. Once the k winning rolls and the $i-k$

neutral rolls, which are independent, are identified, the factors p^k and q^{i-k} account for the probabilities of the individual rolls.

The factors p^k and q^{-k} do not depend on the summation variable i and so may be factored out of the sum, giving

$$P(\text{Win}) = \left(\frac{p}{q}\right)^k \sum_{i=k}^{\infty} \binom{i-1}{k-1} q^i.$$

The online computer algebra system Wolfram Alpha can be used to evaluate this series: For $|q| < 1$, we have

$$\sum_{i=k}^{\infty} \binom{i-1}{k-1} q^i = \left(\frac{q}{1-q}\right)^k,$$

and

$$P(\text{Win}) = \left(\frac{p}{1-q}\right)^k.$$

We can eliminate one variable by substituting $q = \frac{5}{6} - p$ to give

$$P(\text{Win}) = \left(\frac{6p}{6p+1}\right)^k.$$

Example 2.1. For the Repeater bet on the number 10, we have $k = 4$, $p = \frac{1}{12}$, and $q = \frac{3}{4}$. The probability of winning is then $\frac{1}{81} \approx .0123$, giving the casino a 19.75% advantage. ∎

House advantages on this version of the Repeater bet range from 18.37% on 2 or 12 to 21.88% on 3, 11, or the combination bet covering 2, 3, 11, and 12.

An alternate pay table for Repeater calls for the number n to be rolled n times before a 7 is rolled. The payoffs for the numbers 2–6 are unchanged; the new payoffs on 8–12 are shown in Table 2.2.

TABLE 2.2: Repeater bet alternate target numbers and payoff odds [14]

Roll	Target number	Payoff odds
8	8	400 for 1
9	9	2500 for 1
10	10	25,000 for 1
11	11	1,000,000 for 1
12	12	50,000,000 for 1

This Repeater bet on 12 is a wager that 12 will be thrown 12 times before a 7 is thrown once. At a 50 million for 1 payoff, this is among the highest-paying bets proposed for casino games. The probability of winning this bet is

$$\left(\frac{1}{7}\right)^{12} \approx 7.2248 \times 10^{-11},$$

approximately 1 in 13.8 billion, and the house advantage is 99.84%—also among the highest of any wager. If the payoff on this bet was raised to 10 billion for 1, the HA would drop to 27.72%—far more reasonable, but with a nonzero chance of bankrupting the casino.

There are other options, of course. One possibility might be called the "Boxcar Bonus" bet after the slang term for rolling two 6s. To attract some wagers on this longest of shots, a game designer might propose a pay table like Table 2.3, which pays off smaller amounts for fewer than twelve 12s.

TABLE 2.3: Boxcar Bonus payoffs on 12s

Number of 12s	Payoff (for 1)
5	5
6	50
7	500
8	5000
9	50,000
10	500,000
11	5,000,000
12	50,000,000

While this enhanced pay table might attract new players, the notion that they're getting a meaningfully better game is an illusion, as the HA on this bet has only fallen from 99.84% to 99.02%. The probability of winning the smallest prize by rolling five 12s before a 7 is only 5.95×10^{-5}, or 1 chance in 16,807.

2. The probability of winning a particular wager.

If the craps Don't Pass wager is a simple mirror image of the Pass line wager, the player holds an advantage. The casino's edge is restored when a 1-out-of-36 roll, either the 2 or 12, is barred on the come-out roll. Turning this win into a push returns the edge to the house.

Craps using 8-sided dice was proposed by Richard Epstein under the name *Sparc* ("craps" spelled backward) [21, p. 214]. Figure 2.1 displays two such dice, which are called *octahedra*. *Spider Craps* fleshed out Epstein's idea to a fully formed casino game. The game was developed by Jacob Engel in 2011.

A Pass line bet at Spider Craps mimics standard craps: the bet wins on an initial 9 or 15, loses on a 2, 3, or 16, and establishes a point if any other

FIGURE 2.1: Eight-sided dice used in Spider Craps.

number is rolled. The direct Spider Craps analog of the standard Don't Pass bet would be

- Win if the come-out roll is 2 or 3.
- Lose if the come-out roll is 9 or 15.
- Push if the come-out roll is 16, just as a craps Don't Pass bet pushes on an initial roll of 12.
- If a point is established on the come-out roll, reroll until either the point or a 9 is rolled. If the point is rerolled before a 9, Don't Pass loses; if a 9 turns up first, the bet wins.

It turns out that excluding only 1 roll in 64 (the 16) does not go far enough in balancing the game for the casino. Under these rules, we have

$$P(\text{Win at Don't Pass}) = \left(1 - \frac{37,319}{80,080}\right) - \frac{1}{64} \approx .5184.$$

Since the player would have the advantage on this even-money bet, we reject this rule. The rules of the Don't Pass bet need to be changed in a way that gives the casino a small edge—large enough so the house makes a profit, but not so large that players are driven away.

If both the 2 and the 16 are barred on the come-out roll, the probability of winning only falls to .5027, so this is not enough. If instead the 2 and 3 are barred, the new Don't Pass wager wins with probability

$$\left(1 - \frac{37,319}{80,080}\right) - \frac{3}{64} = \frac{156,029}{320,320} \approx .4871.$$

This gives a HA of 2.58%—an edge that is reasonable without being overwhelming. The rules for the come-out roll on a Spider Craps Don't Pass bet are thus:

- Win instantly on a 16.
- Lose immediately on a 9 or 15.
- Push if the come-out roll is 2 or 3.

This rule is easy for Spider Craps dealers to implement. Any other value becomes the point, and the Don't Pass bettor wins if the shooter rolls a 9 before re-rolling it.

3. The payoff odds to the winning gambler.

In designing a wager and selecting the payoff, several criteria are important.

- The bet should be attractive to players. A 55–1 payoff might trigger a "Wow!" response and inspire some table action.

- The rules should be both easy for players to understand and easy for dealers to implement. A satisfactory HA for some wager might be achievable by setting the payoff odds at 28.4 to 1, but that's too complicated for quick game play in a live casino. If, however, the game is played electronically without human dealers, fractional payoff odds become a design option.

- As always, the HA should straddle the line between being lucrative for casinos but not so lucrative that players reject the bet.

Casino baccarat culture allows players to track the outcome of each game dealt from a shoe and look for patterns in an effort to guide their wagering. Some casinos facilitate this superstition by providing scorecards and special two-color ballpoint pens with both red and black ink to make recording results as easy as possible. As baccarat has been shown to be essentially immune to card counting, any patterns that a gambler might identify are almost certainly meaningless [6]. The *World* side bet, approved for play in Washington state casinos, might appeal to a player seeking to capitalize on a run of three winning hands for either the Banker or Player hand [36].

The World bet is made by choosing a side, Player or Banker, and wins if that side wins three straight hands. Unlike in standard baccarat, a tie hand causes all World bets to lose. Variable pay tables allow Washington casinos to customize the HA as management might wish.

On a given hand of baccarat, the 3 outcomes have the probabilities in Table 2.4. Since successive baccarat hands are effectively independent, it follows that

TABLE 2.4: Baccarat hand probabilities [6]

Result	Probability
Banker wins	.4584
Player wins	.4461
Tie	.0955

the probability of winning a World bet on Banker is

$$(.4584)^3 \approx .0963 \approx \frac{1}{10.4}$$

and the probability of winning a World bet on Player is

$$(.4461)^3 \approx .0888 \approx \frac{1}{11.3}.$$

If the Banker World bet pays off at X to 1, the resulting expectation is

$$E = (X) \cdot .0963 + (-1) \cdot .9037.$$

The approved pay tables for the World bet in Washington allow X to be 8, 8.5, or 9; this allows the HA to range from 3.70–13.33%. For the slightly less likely bet on Player, a 10–1 payoff in addition to 8, 8.5, and 9 is permitted. HAs then range from 2.32% to 20.08%.

Of course, these lower odds and higher house edge can be masked by advertising payoffs as "x for 1" instead of "x to 1."

A Bigger Picture

In general, for a given wager paying off at X to 1 with winning probability p, the expected value of a \$1 bet is a function of two variables:

$$E(X,p) = (X) \cdot p + (-1) \cdot (1-p) = Xp - 1 + p = p(X+1) - 1.$$

The wager is fair if

$$p = \frac{1}{X+1},$$

or if

$$X = \frac{1}{p} - 1 = \frac{1-p}{p}.$$

If p is known, as is frequently the case, this expression sets an upper bound on the payoff X while assuring the casino an advantage.

More complex bets which have different payoffs for different outcomes simply incorporate more variables into the expression for E.

Example 2.2. At Big Red Keno in Omaha, Nebraska, the standard Pick 4 game pays off when a ticket catches either 3 or 4 numbers. If the payoffs for these catches are denoted x_3 for 1 and x_4 for 1 respectively, we have

$$E(x_3, x_4) = (x_3) \cdot \frac{\binom{4}{3} \cdot \binom{76}{17}}{\binom{80}{20}} + (x_4) \cdot \frac{\binom{76}{16}}{\binom{80}{20}} - 1 \approx .0432x_3 + .0031x_4 - 1.$$

While the probabilities of winning are fixed, the HA may be manipulated by changing the payoffs. The standard game has $x_3 = 1$ and $x_4 = 180$, which gives a HA of 40.53%. In March 2019, the monthly special raised x_4 to 200, which cut the HA to 34.41%.

If $x_3 = 1$ and $x_4 = 300$, the HA falls to 3.77%, which is close to an even game. The upper bound on X computed above must be adjusted by adding 1 to account for the "for 1" payoff that is standard in keno. If $x_3 = 0$, then the upper limit on x_4 is

$$\frac{1-p}{p} + 1 = \frac{1}{p} = \frac{\binom{80}{20}}{\binom{76}{16}} \approx 326.44.$$

∎

2.3   An Extended Example: Casino Merry-Go-Round

As an illustration of these three design principles, we shall look at *Casino Merry-Go-Round*, a game patented in 1992 that was played with 6 nonstandard 6-sided dice.

1. Choosing the equipment

Most casino games are intentionally easy to learn and to understand. In roulette, the winning number and all of its properties (odd/even and the like) are on display when the ball lands. Blackjack and baccarat require nothing more than simple counting and familiarity with the object of the games. Craps may be a bit more complicated than these other games, but everyone who can count and add can identify the number rolled on the dice.

Casino Merry-Go-Round, played with 6d6, charted a different course. While the essence of the game was simple—bet on whether or not one die or sum of two dice would be bigger than another specified die or pair—the dice were numbered unconventionally and enclosed in a cage, and that cage was spun only by the dealer. This, the patent asserted, would make it more difficult for players to figure out how the dice were numbered and impossible for a controlled shooter to influence how they landed [70].

The 6 dice were two sets of 3 *nontransitive* dice.

Definition 2.1. Three dice A, B, and C are *nontransitive* if they are numbered in such a way that the following are all true:

- A beats B, by showing a higher number when rolled, at least 50% of the time.
- B beats C at least 50% of the time.
- C beats A at least 50% of the time.

Nontransitive dice are certainly counterintuitive: if A is "better" than B in this sense, and B is likewise better than C, we would reasonably expect that A would also be better than C. This definition may be extended to sets of 4 or more dice, as in the following example.

FIGURE 2.2: A set of 4 nontransitive dice.

Example 2.3. Consider the dice shown in Figure 2.2. From left to right, we have the following:

- A: 3,3,3,3,3,3.
- B: 2,2,2,2,6,6.
- C: 1,1,1,5,5,5.
- D: 0,0,4,4,4,4.

Then A beats B $\frac{2}{3}$ of the time (whenever B rolls a 2), B beats C $\frac{2}{3}$ of the time, C beats D $\frac{2}{3}$ of the time, and D beats A $\frac{2}{3}$ of the time. These 4 dice form a nontransitive set. ∎

The four dice A–D above could be used in a lucrative informal game, perhaps at a bar. Bet a friend that you can roll a higher number on one die than he can, and offer him first choice of the dice. Once he has made a selection, you choose the die that beats it $\frac{2}{3}$ of the time. As an even-money wager, this gives you a $33\frac{1}{3}\%$ advantage.

Casino Merry-Go-Round distinguished its dice by color and size [70].

- Big Red: 1,4,4,4,4,4.
- Big White: 2,2,2,5,5,5.
- Big Green: 3,3,3,3,3,6.
- Small Red: 2,2,2,5,5,5.
- Small White: 3,3,3,3,3,6.
- Small Green: 1,4,4,4,4,4.

That the big dice and small dice each comprise a set of 3 nontransitive dice can be seen by direct calculation. The big red die beats the big green die if it shows a 4 while the big green die shows a 3; this has probability

$$\frac{5}{6} \cdot \frac{5}{6} = \frac{25}{36} > \frac{1}{2}.$$

Similarly, the probability that the big green die beats the big white die is

$$\frac{5}{6} \cdot \frac{1}{2} + \frac{1}{6} = \frac{7}{12} > \frac{1}{2},$$

and the probability that the big white die beats the big red die is

$$\frac{1}{2} \cdot \frac{1}{6} + \frac{1}{2} = \frac{7}{12} > \frac{1}{2}.$$

Since the small dice bear the same combinations of numbers as the large dice, they also form a nontransitive set.

The decision to use 6 dice was made in an effort to provide a wide array of betting options and to ensure the casino held an advantage on each bet. Moreover, while there are several other ways to number 3 dice as a nontransitive triple, the inventors thought that restricting the numbers on the dice to the familiar 1–6 would make the game more appealing to players. One simple way to produce another set of nontransitive dice would simply be to start with the 6 dice above and then add a fixed value to each number on each die.

Example 2.4. The following numbering scheme also comprises a nontransitive set of 3 dice [70]:

- A: 1,4,6,6,8,8.
- B: 3,3,3,5,7,10.
- C: 2,2,2,9,9,9.

The probability that A beats B is

$$\frac{1}{6} \cdot \frac{3}{6} + \frac{2}{6} \cdot \frac{4}{6} + \frac{2}{6} \cdot \frac{5}{6} = \frac{21}{36},$$

where the three terms in the sum represent the probabilities, respectively, of A outscoring B when rolling a 4, 6, and 8. The probability that B beats C is

$$\frac{3}{6} \cdot \frac{3}{6} + \frac{1}{6} \cdot \frac{3}{6} + \frac{1}{6} \cdot \frac{3}{6} + \frac{1}{6} \cdot \frac{6}{6} = \frac{21}{36},$$

and the probability that C beats A is

$$\frac{3}{6} \cdot \frac{1}{6} + \frac{3}{6} \cdot \frac{6}{6} = \frac{21}{36}.$$

All 3 probabilities exceed ½. ∎

2. Selecting the wagers

In addition to being nontransitive, the dice used in Casino Merry-Go-Round and the available wagers were carefully chosen so that no ties were possible. Every spin of the six dice in the cage resolved every bet on the layout.

Casino Merry-Go-Round offered two types of basic wagers: one-die bets where players bet on one die to outroll a specified other die, and two-die bets where the sums of two specified dice on each side were compared with the higher total winning.

In setting the permissible wagers for Casino Merry-Go-Round, the game designers sought, of course, to build in a positive HA on every bet. This led to bets that were decided like the informal game described on page 34 following Example 2.3. In a bet comparing two dice, the player can choose which of the six dice he or she wishes to back, but then the casino always gets a die with a greater than 50% chance of beating it.

Example 2.5. The small red die (2,2,2,5,5,5) beats the small green die (1,4,4,4,4,4) with probability $\frac{21}{36}$, but in the bet matching those two dice, the player must take the small green die. A player wishing to bet on the small red die in a one-die matchup must face off against the small white die (3,3,3,3,3,6), and only has probability $\frac{15}{36}$ of winning.   ■

The 12 bets available to the gambler in Casino Merry-Go-Round are listed in Table 2.5.

TABLE 2.5: Casino Merry-Go-Round bets [70]

Wager	Player's die/dice	Casino's die/dice
A	Both red dice	Both green dice
B	Both white dice	Both red dice
C	Both green dice	Both white dice
D	Big white	Big green
E	Big green	Big red
F	Big red	Big white
G	Big red + Small green	Big green + Small white
H	Big green + Small white	Big white + Small red
I	Big white + Small red	Big red + Small green
J	Small white	Small green
K	Small red	Small white
L	Small green	Small red

The following types of bets are excluded:

- Faceoffs between 2 identically numbered dice, for example: big red and small green. These represent both fair bets and bets where a tie is possible, and those are both outcomes that the casino wishes to avoid.

- Bets where the player would have an advantage, as when any bet in Table 2.5 is reversed.

Consider bet A, where the player has the red sum and the house has the green sum. We have the outcomes for the dice shown in Table 2.6, where X denotes the red sum and Y the green sum. Inspection of the table shows that no ties are possible. Table 2.6 leads to the PDFs shown for both X and Y.

Since the player wins this bet if $X > Y$, the probability of winning is

$$\frac{1}{2} \cdot \frac{5}{36} + \frac{5}{12} \cdot \frac{31}{36} = \frac{185}{432} \approx .4282 < .5.$$

TABLE 2.6: Casino Merry-Go-Round: Outcomes and PDFs when the red dice (X) and green dice (Y) are rolled (wager A)

X ⚀			⚄			Y ⚀	⚁				
3	3	3	6	6	6	⚄	4	7	7	7	7	7
6	6	6	9	9	9		4	7	7	7	7	7
6	6	6	9	9	9		4	7	7	7	7	7
6	6	6	9	9	9		4	7	7	7	7	7
6	6	6	9	9	9		4	7	7	7	7	7
6	6	6	9	9	9	⚅	7	10	10	10	10	10

x	$P(X = x)$	y	$P(Y = y)$
3	$\dfrac{1}{12}$	4	$\dfrac{5}{36}$
6	$\dfrac{1}{2}$	7	$\dfrac{26}{36}$
9	$\dfrac{5}{12}$	10	$\dfrac{5}{36}$

A player seeking more action than is available on a single bet may also choose from 10 groups of 3 of the wagers from Table 2.5. In addition to providing increased betting choices, these composite bets involve more of the dice, and thus greater action, than wagers A–L provide. No matter how the dice should fall, every group has been designed to guarantee that the player will win either 1 or 2 of their 3 wagers. The 12 groups are listed in Table 2.7.

TABLE 2.7: Casino Merry-Go-Round group bets [70]

Group	# of dice	Group	# of dice
ABC	6	BFK	4
DEF	3	CDJ	4
GHI	6	GEJ	4
JKL	3	HDK	4
AEL	4	IFL	4

Example 2.6. A player selecting the ABC bet makes bets A, B, and C: simultaneously betting on the red sum to beat the green sum, the white sum to beat the red sum, and the green sum to beat the white sum. As a result, all 6 dice in the cage are of interest.

As Table 2.6 shows, the red sum can only be 3, 6, or 9 and the green sum is restricted to 4, 7, and 10. Setting up the PDF for the white sum shows that it can only be 5, 8, or 11—so once again, a Casino Merry-Go-Round bet admits no ties.   ∎

3. Setting the payoffs

If paid at 1–1, wager A carries an expected value of

$$E = (1) \cdot \frac{185}{432} + (-1) \cdot \frac{247}{432} = -\frac{62}{432},$$

and a house advantage of 14.35%. Depending on the intentions of game designers and possibly on the other available wagers, this may be seen as too high. A high initial HA may be adjusted by improving the payoff odds. Using the formula for $E(X, p)$ on page 32, we can show that the HA for a \$1 bet on wager A paying X to 1 is

$$1 - \frac{185}{432} \cdot (X + 1).$$

With $X = 2$, the HA is negative, so a more player-friendly wager will have payoff odds between 1–1 and 2–1.

If we desire a 5% house edge, we must solve the equation

$$1 - \frac{185}{432} \cdot (X + 1) = .05$$

for X, which results in $X = \frac{1127}{925} \approx 1.2183$.

Odds of 1.2183–1 are far too complicated for efficient gameplay, but rounding to 1.2–1, or 6–5, gives a workable game with a HA of 5.79%. Table 2.8 shows the house advantages for a range of payoff options between 1–1 and 2–1.

TABLE 2.8: Casino Merry-Go-Round: Payoff options for wager A

X	Payoff odds	HA
1.000	1–1	14.35%
1.$\overline{111}$	10–9	10.07%
1.125	9–8	9.00%
1.$\overline{166}$	7–6	7.21%
1.200	6–5	5.79%
1.250	5–4	3.65%
1.$\overline{333}$	4–3	−0.07%
1.500	3–2	−7.06%

The Casino Merry-Go-Round patent recommends a 9–8 payoff ($X = 1.125$) on wager A, which gives the casino a 9.00% edge [70].

Example 2.7. Casino Merry-Go-Round wager E is a simple proposition where the player has the big green die (3,3,3,3,3,6) and the casino the big red die (1,4,4,4,4,4).

The player's winning probability is

$$\frac{5}{6} \cdot \frac{1}{6} + \frac{1}{6} \cdot \frac{6}{6} = \frac{11}{36},$$

slightly less than $\frac{1}{3}$. Paying a winning bet at 2–1 gives an expected value of

$$E = (2) \cdot \frac{11}{36} + (-1) \cdot \frac{25}{36} = -\frac{1}{12} \approx -.0833.$$

The casino's edge is then 8.33%. This wager is an example where a simple integer-to-1 payoff leads to an acceptable HA. ∎

The recommended payoffs for the 12 basic Casino Merry-Go-Round wagers are shown in Table 2.9. Several of the payoffs are of the form "*x* to 8"; while this

TABLE 2.9: Casino Merry-Go-Round payoffs [70]

Wager	Player dice	Casino dice	Payoff
A	Both red	Both green	9–8
B	Both white	Both red	5–4
C	Both green	Both white	9–8
D	Big white	Big green	5–4
E	Big green	Big red	2–1
F	Big red	Big white	5–4
G	Big red + Small green	Big green + Small white	7–8
H	Big green + Small white	Big white + Small red	5–4
I	Big white + Small red	Big red + Small green	5–4
J	Small white	Small green	2–1
K	Small red	Small white	2–1
L	Small green	Small red	5–4

may be a sensible mathematical choice, its effect on gameplay deserves careful investigation. If the game attracts high rollers or many Asian gamblers for whom the number 8 is considered especially lucky, an $8 minimum bet might be a reasonable rule. If players are accustomed to wagering smaller amounts, then paying exact winnings might result in too many fractions. Calculation of these payoffs would slow down the game considerably. If the payoffs were regularly rounded in the casino's favor, as is the case with traditional casino table games, players might object.

Example 2.8. If a player were to place a winning $5 wager on bet C, the 9–8 odds call for a payoff of $5.625. In the likely event that the Casino Merry-Go-Round table is not supplied with coins or chips valued at less than a dollar, the player would receive only \$5, so the effective odds would drop to 1–1. Standard arithmetic rounding rules call for rounding \$5.625 up to \$6, but casinos don't round that way.

A 6–5 payoff on bet C would result in a house edge of 5.79%, which would make C the best available wager.   ∎

Another option would be to set the game's fundamental currency unit at 25¢, so that an 8-unit bet would correspond to \$2. This might be accommodated through the use of special casino chips (called "checks" in this instance) used only at the Casino Merry-Go-Round table, as is done for roulette. Players could buy into the game for the amount of their choice and be issued these 25¢ chips in a color that only they use, to make it easier to keep everyone's bets straight on the layout.

Example 2.9. In Table 2.9, wager G attracts special attention due to its less than even-money payoff of 7–8. The player has the sum of the big red and small green dice (X), while the casino faces off with the big green and small white dice (Y). The outcomes of these rolls are charted in Table 2.10, where the big dice label the rows and the small dice the columns.

TABLE 2.10: Casino Merry-Go-Round: Outcomes and PDFs for wager G: Big red + Small green vs. Big green + Small white

X						Y					
2	5	5	5	5	5	6	6	6	6	6	9
5	8	8	8	8	8	6	6	6	6	6	9
5	8	8	8	8	8	6	6	6	6	6	9
5	8	8	8	8	8	6	6	6	6	6	9
5	8	8	8	8	8	6	6	6	6	6	9
5	8	8	8	8	8	9	9	9	9	9	12

x	$P(X = x)$	y	$P(Y = y)$
2	$\dfrac{1}{36}$	6	$\dfrac{25}{36}$
5	$\dfrac{10}{36}$	9	$\dfrac{10}{36}$
8	$\dfrac{25}{36}$	12	$\dfrac{1}{36}$

The only way that players win bet G is when their dice show an 8 and the casino's dice roll a 6; the probability of this event is $\frac{625}{1296}$. Since this is less than ½, we might reasonably ask why the bet pays 7–8 instead of 1–1. Using

7–8 odds, the expected value of an \$8 bet is

$$E = (7) \cdot \frac{625}{1296} + (-8) \cdot \frac{671}{1296} = -\frac{993}{1296} \approx -\$.7662.$$

Dividing by \$8 gives a casino edge of 9.58%. ∎

Paying bet G at even money would reduce the HA to 3.55%. While not out of line for a carnival game, there is an additional matter to consider here. If a game has a collection of similar-looking bets, it is better for game balance if their HAs are all fairly close to one another. American roulette is a good example of balanced bets, with every bet except the basket bet carrying a 5.26% HA. In an unbalanced game where one bet, say wager G paying even money, is substantially more player-friendly than the others, we expect that action will shift over time, as players learn more about the game, to favor that option. The net result will be a lower overall profit for the casino as players eschew higher-risk bets for those with similar payoffs at less unfavorable odds.

Table 2.11 shows that the HAs for the 12 basic Casino Merry-Go-Round bets all fall in the relatively narrow range from 6.25–9.58%.

TABLE 2.11: Casino Merry-Go-Round house advantages [70]

Wager	HA	Wager	HA
A	9.00%	G	9.58%
B	7.81%	H	7.81%
C	9.00%	I	7.81%
D	6.25%	J	8.33%
E	8.33%	K	6.25%
F	6.25%	L	6.25%

Example 2.10. Nontransitive dice are certainly paradoxical, but the paradoxes don't end with the dice. Examination of Tables 2.9 and 2.11 reveals the following chain of advantages:

- Bet K carries a HA of 6.25%. This is the edge that the small white die has over the small red die.

- Bet D shows that the large green die has a 6.25% advantage over the large white die.

- However, when we combine the 4 dice into 2 pairs to form bet H, the *sum* of the small white and large green dice is the underdog against the sum of the small red and large white dice, with the casino enjoying a 7.81% edge—greater than the edge from either bet K or D.

∎

For the ten group bets listed in Table 2.7, the effective HA of the group is the mean of the HAs of the wagers comprising the group, assuming that an equal wager is made on each option. Consider bet ABC, which involves all 6 dice in same-color pairs. Since there are 3 outcomes for each of the 3 pairs of dice, the probability of a particular outcome on a given same-color pair of dice can be expressed as a quadratic function of the sum. Denoting this sum by x, we have

$$\text{Red and White: } F(x) = -\frac{x^2}{36} + \frac{7x}{18} - \frac{5}{6}$$

and

$$\text{Green: } G(x) = -\frac{7x^2}{108} + \frac{49x}{54} - \frac{265}{108}.$$

Since the 3 pairs of dice are independent, the three functions can be multiplied together to get the probability of any specified outcome: If the sums of the red, white, and green dice are denoted respectively by r, w, and g, we have

$$P(r, w, g) = F(r) \cdot F(w) \cdot G(g).$$

Wager ABC has 27 different outcomes with varying probabilities listed in Table 2.12.

By adding, we find that the probability of winning 1 of the 3 bets is .7338 and the probability of winning 2 out of 3 is .2662. The amount won or lost depends on which bets are won. Whether by summing probabilities in Table 2.12 or by taking the mean of the HAs of wagers A, B, and C, we find that the HA of ABC is 8.60%.

Though it may have seemed like a good thematic fit for the Circus Circus casinos in Las Vegas and Reno, Casino Merry-Go-Round was not a successful carnival game.

2.4   The Role of Variance

The house advantage doesn't tell the entire story of a carnival game, though. Another significant statistic is the *variance* of a random variable, which measures the deviation of the data from the mean or expected value. A very large bet carrying a low HA might nonetheless be lucrative for a casino if the variance is high; this leads to a steady stream of income for the casino until the players run out of money. This happens more quickly, on the average, in a game with high variance since more money is at risk on every decision and the casino has a greater cash reserve than any gambler.

Example 2.11. Wagering \$1 on the number 4 at roulette for 10 consecutive spins leads to a binomial PDF for the winnings X, with parameters $n = 10$

TABLE 2.12: Casino Merry-Go-Round: ABC wager

Roll			Probabilities			Bets	Win	Payoffs			
r	w	g	$P(r)$	$P(w)$	$P(g)$	Won	Prob.	A	B	C	Sum
3	5	4	.0833	.4167	.1389	B	.0048	−8	10	−8	−6
3	5	7	.0833	.4167	.7222	BC	.0251	−8	10	9	11
3	5	10	.0833	.4167	.1389	BC	.0048	−8	10	9	11
3	8	4	.0833	.5000	.1389	B	.0058	−8	10	−8	−6
3	8	7	.0833	.5000	.7222	B	.0301	−8	10	−8	−6
3	8	10	.0833	.5000	.1389	BC	.0058	−8	10	9	11
3	11	4	.0833	.0833	.1389	B	.0010	−8	10	−8	−6
3	11	7	.0833	.0833	.7222	B	.0050	−8	10	−8	−6
3	11	10	.0833	.0833	.1389	B	.0010	−8	10	−8	−6
6	5	4	.5000	.4167	.1389	A	.0289	9	−8	−8	−7
6	5	7	.5000	.4167	.7222	C	.1505	−8	−8	9	−7
6	5	10	.5000	.4167	.1389	C	.0289	−8	−8	9	−7
6	8	4	.5000	.5000	.1389	AB	.0347	9	10	−8	11
6	8	7	.5000	.5000	.7222	B	.1806	−8	10	−8	−6
6	8	10	.5000	.5000	.1389	BC	.0347	−8	10	9	11
6	11	4	.5000	.0833	.1389	AB	.0058	9	10	−8	11
6	11	7	.5000	.0833	.7222	B	.0301	−8	10	−8	−6
6	11	10	.5000	.0833	.1389	B	.0058	−8	10	−8	−6
9	5	4	.4167	.4167	.1389	A	.0241	9	−8	−8	−7
9	5	7	.4167	.4167	.7222	AC	.1254	9	−8	9	10
9	5	10	.4167	.4167	.1389	C	.0241	−8	−8	9	−7
9	8	4	.4167	.5000	.1389	A	.0289	9	−8	−8	−7
9	8	7	.4167	.5000	.7222	A	.1505	9	−8	−8	−7
9	8	10	.4167	.5000	.1389	C	.0289	−8	−8	9	−7
9	11	4	.4167	.0833	.1389	AB	.0048	9	10	−8	11
9	11	7	.4167	.0833	.7222	AB	.0251	9	10	−8	11
9	11	10	.4167	.0833	.1389	B	.0048	−8	10	−8	−6

and $p = \frac{1}{38}$. The 11 possible values of X run from −$10 to $350 in steps of $36.

If that $10 is wagered on 4 at a single spin, there are only two possible outcomes: −$10 and $350. In each case, the *mean* outcome for the $10 worth of bets is −52.6¢, but the outcome of the single bet will typically be farther from that mean than the outcome of the ten consecutive smaller spins. The variance of X measures this distance. ∎

Definition 2.2. Let X be a random variable with expectation $\mu = E(X)$. The *variance* of X, denoted Var(X) or σ^2, where σ is the Greek lowercase letter sigma, is

$$\sigma^2 = \sum_x \left[(x - \mu)^2 \cdot P(X = x) \right].$$

This can be rearranged to the equivalent expressions

$$\sigma^2 = E(X^2) - E(X)^2$$
$$= E(X^2) - \mu^2.$$

Since the variance is the sum of nonnegative terms, it is always greater than or equal to 0, and is equal to zero if and only if X can take on only one value, with probability 1.

Definition 2.3. The *standard deviation* σ of X is the square root of the variance.

As the square root of a nonnegative quantity, σ, like σ^2, is also nonnegative.

The units of variance are the *square* of the units of X, so if X is measured in dollars, σ^2 is measured in dollars squared. The standard deviation (abbreviated SD) measures the "typical" difference of a data point from the mean, and so has the same units as X.

Example 2.12. Continuing Example 2.11, the variance of ten consecutive roulette \$1 bets on a single number is 332.08 dollars squared. The SD of this sequence of bets is \$18.22.

The variance of a single \$10 bet on that number is ten times greater: 3370.28 dollars squared. The corresponding SD is \$57.23. This greater SD means that the results of a single \$10 bet are more variable than the outcome of 10 individual \$1 wagers. The typical outcome of a \$10 bet is \$57.23 away from the mean result, while the typical outcome of ten \$1 bets is only \$18.22 from its mean.

∎

Consider a baccarat bet on Player, where the probability distribution of a \$1 bet is [6]:

x	−1	1	0
$P(X = x)$	.4584	.4461	.0955

This bet has a HA of 1.23% and a variance of .9043 dollars squared. If the player instead bets \$100,000, the PDF of that bet is

x	−100,000	100,000	0
$P(X = x)$	.4584	.4461	.0955

and while the HA remains 1.23%, the variance is now 9,043,487,100—over 9 billion dollars squared, and 10 billion times greater than the variance for the \$1 bet. While the expected value of this \$100,000 bet might be only −\$1230, which is the same 1.23% of the wager as applies to the \$1 bet, repeated wagers at this level are subject to wild swings about the mean.

Measured in absolute dollar amounts rather than as a percentage of the wager, these swings will bankrupt the player before ruining any casino that is capitalized enough to justify allowing its patrons to place such a large bet. A player who loses 10 more bets than she wins is down $1,000,000, and while this may or may not be a financial hardship for her, a well-run casino that loses $1,000,000 to a single player is certainly able to focus on the long term and feel little or no short-term stress. Indeed, a casino accustomed to big action like this is liable to consider a million-dollar player win as a temporary loan to that player, secure in the knowledge that, over time, the gambler will lose that money right back.

To a game designer, a bet with high variance can justify a low HA, provided that it can attract high rollers who will place the large bets essential to making the mathematics work out in practice.

Example 2.13. The Pass line bet at craps offers an attractive probability of winning, .4929, and a very low house advantage of 1.41% [6]. Casinos who offer free odds bets backing up a Pass wager give players a fair bet, since free odds bets are paid off at their true odds with no additional HA. This combination of wagers then faces a lower HA in total, but the variance rises since more money is ultimately at risk.

Suppose that a player makes a $10 Pass line bet and the point is 8. Since the come-out roll has not resolved the bet, the probability of winning has dropped to $\frac{5}{11} \approx .4545$ and the HA of this $10 bet is 9.09%. The player then backs this Pass bet up with a 5× odds bet of $50, which pays off at 6–5 if an 8 is rolled before a 7. There is $60 at risk, with the following PDF for the player's winnings X:

Outcome	Probability
+70	.4545
−60	.5454

The HA on the combined bet has dropped to 1.52%.

This improved HA has come at the price of a greater risk, though, as the player now has $60 on the line at a disadvantage rather than $10. A look at the variance will illustrate the difference. For the original $10 wager, we have

$$\sigma^2 = (10)^2 \cdot .4929 + (-10)^2 \cdot .5071 - (-1.414)^2 \approx 96.00,$$

while the variance of the wager with free 5× odds is

$$\sigma^2 = (70)^2 \cdot \frac{5}{11} + (-60)^2 \cdot \frac{6}{11} - (-1.818)^2 \approx 4187.60.$$

Taking square roots, we see that the standard deviation of this bet is $9.80 without odds and $64.71 with the odds bet. Making a free odds bet may make the combined bet a better gamble, but this comes at the cost of placing more money at risk and increased volatility. ∎

2.5   Exercises

Solutions begin on page 285.

1. Find the HA for a $1 wager on a version of Red Dragon (page 21) played with

a. 14 dice.

b. 16 dice.

Casino Merry-Go-Round

2. Can a pair of nonstandard d6s, numbered so that no ties are possible, be nontransitive? Explain why or why not.

3. Find a value of x, where x is an integer in the range $1 \leqslant x \leqslant 10$, such that the set of dice

- A: 1,4,4,4,4,x.
- B: 3,3,3,3,5,8.
- C: 2,2,2,7,7,7.

is nontransitive.

4. Investigate the effect on the HA of changing the payoff odds on the following Casino Merry-Go-Round bets, and find which payoff odds using whole numbers less than or equal to 10 produce the smallest positive HA.

a. Wager C.

b. Wager I.

c. Wager J.

A 3d4 Game

The following game played with three 4-sided dice was described in a 1987 US patent [32]:

> Two 4-sided dice are white, with each side bearing a number from 1–4. The third d4 is also numbered 1–4, and each side is colored differently. The game is played in two stages.
>
> 1. Players may make a variety of bets on the numbers showing on the white dice or on all 3 dice when thrown.

    A. The numbers on the white dice: 1–1, 1–2, 1–3, 1–4, 2–2, 2–3, 2–4, 3–3, 3–4, or 4–4.

    B. The sum of all 3 dice: odd or even.

    C. The occurrence of a hard 3, 6, 9, or 12, where a hard total arises from all 3 dice showing the same number.

    D. The sum of all 3 dice: 5 through 10.

    E. The 3 numbers appearing on all 3 dice: 16 combinations (triples are excluded) from 1–1–2 up through 3–4–4.

One of the gamblers rolls the two white dice and the bets in category A are resolved.

2. Players may then make separate bets on the color and number rolled on the multicolored die. This third die is rolled, and all bets are settled.

Standard d4 game dice are regular tetrahedra, with 4 congruent equilateral triangular sides. These dice do not necessarily roll and bounce in a reliably random manner when tossed. With that in mind, the game designer suggested 4-sided dice made up of 4 mutually tangent spheres, each labeled with a different number. The spheres comprising the multicolored die would each be a different color—perhaps, following the colors of standard pool balls, 1 would be yellow, 2 blue, 3 red, and 4 purple. Figure 2.3 shows this die.

FIGURE 2.3: A non-tetrahedral d4 made from 4 mutually tangent spheres.

Solve the following problems about this game.

5. The category A bets, on the two numbers shown on the white dice, can be separated into two types: bets on doubles and bets on non-doubles.

a. Suppose that a bet on a specified double pays off at x to 1. Find an expression for the expected value of a \$1 wager as a function of x, and use it to find an integer value of x with the smallest positive HA.

b. Repeat part a for a bet on a non-double combination paying x to 1.

c. This game could also be marketed for home play. In such a game, it is desirable that there be no HA, so that players are on even footing. Find the appropriate value for x on each of these 2 bets so that the wagers are fair.

6. Find an appropriate payoff for the hard way 3, 6, 9, and 12 bets (category C) so that the HA falls between 2% and 6%.

7. For the category D wagers on the sum of all 3 dice, compute the probability of winning a bet on each number from 5 through 10.

8. Category E wagers on the combinations appearing on all 3 dice admit 2 mathematically different bets. For the following combinations of numbers appearing, find the payoff odds that yield a fair bet.

a. $x - y - z$, where x, y, and z denote different values in the range 1–4.

b. $x - x - y$, where $x \neq y$.

9. If a bet on the color or number of the multicolored die pays off at 5–2, find the HA.

10. Consider a variation of this game where the d4s are replaced by d6s. Repeat Exercise 5 for this game.

Beat the House

Beat the House is played with 3d6, one colored red and two colored white. The red die is the players' die, while the white dice are the house's. All 3 dice are rolled together, and the players win if the number on their die exceeds the sum of the house dice [60].

11. Disregarding ties for the moment, find the probability of a player win.

If the player die ties the house dice, players may surrender their bets, forfeiting half their wager, or proceed to a second roll. The 3 dice are rolled again and the players' die compared to the house sum as on the first roll. If the second roll results in a tie or if the players' die beats the house dice, the players win.

12. Find the probability of a tie on the first roll.

13. What is the probability of a player win in either 1 or 2 rolls?

14. Suppose that all winning bets on the players' dice paid off at fixed odds of x–1. Find the integer value of x that gives the smallest positive HA.

15. An alternate pay table for Beat the House pays increasing odds as the margin of the players' victory increases, with the top payoff going to a player ⚅ against a dealer ⚀ . Table 2.13 shows the payoff odds.

TABLE 2.13: Pay table for Beat the House based on players' victory margin [60]

Margin of victory	Payoff odds
0	Go to second round
1	3–1
2	5–1
3	10–1
4	50–1

If the player and dealer tie on the first round, action on this bet is deferred to the second round. Should the player beat the house on this roll, the payoff is 20 times the amount wagered [60]. Find the house advantage of the game using this pay table.

Chapter 3

Wheel and Ball Games

As noted on page 20, the adage "The easier a bet is to understand, the higher the house advantage" is a simple summary of house advantage that proves remarkably reliable in the field. Some of the simplest casino games are played with wheels and balls of various sorts, including the familiar roulette, keno, bingo, and Big Six wheel. These games typically require very little skill and are simply a matter of picking a number or numbers and hoping they turn up when a wheel is spun or some ping-pong balls are randomly shaken and drawn. They also tend to have high house edges.

The most famous wheel-and-ball casino game is *roulette*, in which a wheel divided into multiple pockets—37 in European roulette, 38 in American roulette—is spun and a small ball is released into the wheel. As the wheel and ball slow down, the ball bounces around, eventually settling into one of the pockets. The pocket determines a winning number and color.

That's certainly simple, and the HA of 5.26% on every American roulette wager except the 5-number basket bet rates pretty high among casino table games. European roulette fares better, with a constant HA of 2.70%. Some casinos offer both games side by side, but require a higher minimum bet—perhaps $25 rather than $5—on the European tables in an effort to guarantee similar income from the two games even as the casino's edge differs.

3.1 Stakes and Odds

Stakes and Odds was a series of roulette-based games devised to study human decision-making processes, and was offered at the Four Queens Casino in Las Vegas for a short time in the late 1960s. These games all had a house advantage of 0% when played properly, and while the player could give the casino an edge by his or her choices, there was no way to manipulate the rules to gain an edge for the player. As this was a research game, only one player could play it at a time, in a session that lasted from two to three hours. Stakes and Odds games were played with a PDP-7 computer—very novel for 1969—and an American roulette wheel. The computer was only used to present information and keep track of wagers; all bets were played out on the wheel.

The Stakes and Odds experiment ran for ten weeks with a variety of different games offered [74]. The best-known Stakes and Odds game was Game #1, which was played in two stages. In the first, the computer presented the player with four possible bets, two labeled "Good" and two "Bad," and the player selected one of each type to be played out on the wheel. As presented to the player, a Stakes and Odds bet looked like this, with X, Y, and Z positive integers:

$$\boxed{\begin{array}{c} \text{Pick } X \\ \text{Yours—Win } Y \\ \text{Others—Lose } Z \end{array}}$$

A bet displayed like this was interpreted as follows: The player would use the computer to select X horizontal rows of three numbers on the roulette layout as his or her numbers for one spin of the wheel. There are 12 rows from which to choose; the layout is shown in Figure 3.1.

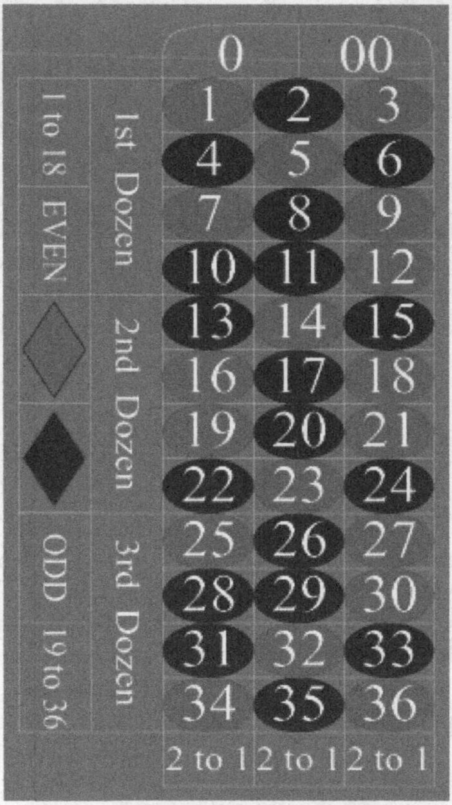

FIGURE 3.1: American roulette layout [25].

0 and 00 were not used in Stakes and Odds, but the Four Queens did not make the considerable investment in a 36-pocket wheel for such a limited experiment. Stakes and Odds used a standard American roulette wheel; if either zero came up during the game, it was disregarded and the wheel was respun. If one of the player's numbers was spun, the payoff was Y chips; if a losing number came up, he or she lost Z chips. Chips were assigned a value by the player at buy-in and could be valued anywhere from 5¢ to $5, with a required initial purchase of 250 chips. In the duration of the game's run at the Four Queens, no player valued their chips at more than 25 cents [74].

In Game #1, a bet would be considered "good" if it afforded the player either a large number of roulette numbers (X) or a high payoff (Y); a "bad" bet arose from a high loss (Z) when the chosen numbers were not spun. The bets were classified into "P" bets, which were characterized by a high probability of winning (good bets with a large X) or losing (bad bets with a small X) and "$" bets, which had larger amounts at stake to win (good bets, with $Y > Z$) or lose (bad bets, $Y < Z$). The specific bets available are listed in Table 3.1 [48]. Whether players opted for P bets or $ bets among the two good and two bad bets presented was the focus of the study.

TABLE 3.1: Stakes and Odds bets [48]

Positive expectation (good)						Negative expectation (bad)					
P bets			$ bets			P bets			$ bets		
X	Y	Z	X	Y	Z	X	Y	Z	X	Y	Z
7	17	7	2	97	11	5	7	17	10	11	97
10	9	3	3	91	21	2	3	9	9	21	91
9	10	2	3	73	15	3	2	10	9	15	73
8	16	11	3	94	22	4	11	16	9	22	94
11	12	24	2	79	5	1	24	12	10	5	79
11	10	2	5	65	31	1	2	10	7	31	65
10	16	2	5	48	12	2	2	16	7	12	48
9	18	2	3	85	11	3	2	18	9	11	85
10	20	10	5	64	20	2	10	20	7	20	64
8	30	15	4	95	25	4	15	30	8	25	95

Example 3.1. There are two $ bets on the first line of Table 3.1. The good bet has $X = 2, Y = 97$, and $Z = 11$. The player chooses 6 numbers and wins 97 chips if one of them is spun while losing 11 if one of the other 30 numbers turns up. The expected value is

$$E = (97) \cdot \frac{6}{36} + (-11) \cdot \frac{30}{36} = \frac{252}{36} = 7.$$

For the bad bet, we have $X = 10, Y = 11$, and $Z = 97$. This bet is the reverse of the first bet, in that the player is now betting on 30 numbers and risking 97 chips while hoping to win 11. Its expected value is –7 chips. ∎

The expected value of one of these bets is a simple function of $X, Y,$ and Z—whether this "simple" expression could be easily evaluated by a player under casino conditions was a different matter. Since the zeroes were not used, the expected value of a bet above is

$$E(X,Y,Z) = \underbrace{(Y) \cdot \frac{3X}{36}}_{\text{Win}} + \underbrace{(-Z) \cdot \frac{36 - 3X}{36}}_{\text{Lose}} = \frac{XY}{12} + Z \cdot \left(\frac{X}{12} - 1 \right).$$

In considering this simplified expression, it is important to note that X must be an integer in the range $1 \leqslant X \leqslant 11$, hence the factor in parentheses by which Z is multiplied, representing a loss, is negative.

In the first round of actual game play, the player was presented with a choice between two good bets and two bad bets, all with the same expected value. The two good bets were one P bet and one $ bet, as were the two bad bets.

Example 3.2. Consider the following four Stakes and Odds bets, which represent one possible collection of wagers presented to a player:

Pick 9	Pick 3
Yours—Win 10	Yours—Win 73
Others—Lose 2	Others—Lose 15
Pick 4	Pick 9
Yours—Win 11	Yours—Win 22
Others—Lose 16	Others—Lose 94

The top row contains the good bets and the bottom row contains the bad bets. Applying the expectation formula derived above, with 1 chip = $1, shows that the good bets each have an expectation of $7 and the bad bets have an expectation of –$7. Picking one good and one bad bet would have a combined expectation of $0 and thus a 0% house edge.   ∎

Which bet from each pair should you choose? Since the expectations are identical, the question ought to come down to your comfort level with risk: are you more interested in a high probability of winning/low probability of losing (P bet) or a higher pair of payoffs ($ bet)?

In the second stage of Game #1, players were presented with all 40 of the previous bets from Table 3.1 in succession. For each good bet, the player was required to state a price—in chips—for which he or she would sell the bet in lieu of playing it. For each bad bet, the player was asked to declare a price that he or she would pay the dealer in order not to have to play the bet out on the wheel. The bets were not labeled as good or bad in stage 2; this assessment was left to the player. Prices that exceeded the potential win (for good bets) or the potential loss (for bad bets) were rejected by the dealer, and a new price was then solicited.

Example 3.3. The bet

> Pick 10
> Yours—Win 16
> Others—Lose 2

is a good bet, with a high probability of winning and a small amount lost if
it doesn't win. While you can win \$16 with probability $\frac{10}{12}$, you might value
the bet at \$13—its expected value—instead of playing it out, and name this
as your price. The casino would reject any bid over \$16. ∎

In either case, the dealer then spun the roulette wheel to generate a random
counteroffer C based on the following formula, with E the expectation of the
bet and r the number spun (again with 0 and 00 disregarded):

$$C = \begin{cases} |E| + \dfrac{r}{2} - 10, & r \text{ even} \\[2ex] |E| + \dfrac{r+1}{2} - 10, & r \text{ odd} \end{cases}.$$

Given a balanced roulette wheel, C ranged from $|E| - 9$, if a 1 or 2 was
spun, to $|E| + 8$, if the wheel came up 35 or 36. If the counteroffer equaled
or exceeded the player's price, then the bet was sold for the amount of the
counteroffer and was not played. If the counteroffer was lower than the player's
price, the bet was played out on the roulette wheel [48].

Example 3.4. Continuing Example 3.3, if the dealer spun a 36, the counteroffer would be \$21. Since $21 > 13$, the player would receive \$21 without
having to play the bet. If the dealer's spin turned up an 18, then $C = \$12$,
and since this is less than the player's price, the bet would be played out, not
sold. ∎

Example 3.5. Consider the bad bet

> Pick 2
> Yours—Win 2
> Others—Lose 16

The expected value of this bet is –\$13. Again, you might choose to price
the bet at –\$13 and simply offer to pay the expectation to avoid risking a loss
of 3 more dollars, which has a probability of $\frac{5}{6}$. If the dealer spun a 2, his
counteroffer would be \$4, and you would be able to escape this bet for \$4,
which is less than your initial price. If the spin landed on 26, the counteroffer
would be \$16. Since this is more than your price, the bet will be played out
on the wheel.

A bet of \$17 or more would be unacceptable to the casino as well as
unwise—why pay \$17 to get out of a bet that, if lost, only costs \$16? ∎

There were eighteen possible values for C, each equally likely with probability $\frac{1}{18}$. The expected value of each transaction—whether the bet was sold or played out—depended on the player's price, but was maximized when the player named the absolute expectation E of the bet as his or her price—either to be paid by the player or the dealer, depending on whether the bet was bad or good. In this case, the bet would be played out half the time, returning E on average, and sold the remaining half, with a return ranging from E to $E + 8$. At a price of E, the expectation to the player was

$$\underbrace{(E) \cdot \left(\frac{1}{2}\right)}_{\text{Played}} + \underbrace{\sum_{k=0}^{8} \left[(E + k) \cdot \frac{1}{18}\right]}_{\text{Sold}} = \frac{18E + 36}{18} = E + 2,$$

a positive expectation which was balanced by charging the player 2 chips for each wager—thus this second stage of Stakes and Odds also had no house advantage if played correctly.

Example 3.6. If the player's price was $E + 4$, then the bet would be sold if the counteroffer was $E + 4$ or higher ($\frac{5}{18}$ of the time), and played if the wheel generated a counteroffer less than $E + 4$. The expectation with this bet would be

$$(E) \cdot \left(\frac{13}{18}\right) + \sum_{k=4}^{8} \left[(E + k) \cdot \frac{1}{18}\right] = \frac{18E + 30}{18} = E + \frac{5}{3} < E + 2.$$

∎

Evaluating the expectation on the fly with the distracting lights and sounds that are present on any casino floor would likely be a challenge to all but the most mathematically adept players.

Did the players play Game #1 correctly? There were 53 completed games during this game's run; the largest recorded win was \$83.30, the biggest loss was \$82.75, and the mean outcome was –\$2.36. No one got terribly rich off Game #1; the relatively small mean loss may be interpreted as a minor fluctuation from random chance—an expected outcome from a game with zero house edge, properly played. However, in the words of the researchers, "The widespread belief that decision makers can behave optimally when it is worthwhile for them to do so gained no support from this study" [48]. One possible explanation offered for this deviation from best possible behavior was that players deliberately overpriced good bets in order to enjoy the thrill of playing out a bet on the wheel.

Several other games were part of the Stakes and Odds experiment, but were not as completely analyzed in the professional psychology literature. Game #9 noted that a traditional roulette player has the option of choosing both the stakes and odds of their wagers: the stakes by deciding how much money to bet, and the odds by selecting how many numbers to cover with a wager. This game allowed players to choose only the stakes or the odds; the other component was computer-selected.

Fourteen bets were offered: 7 in which the player chose the odds and 7 where the player set the stakes. For the Odds bets, the computer first set the stakes at 25¢, 50¢, $1, $2, $5, $10, or $25. Players could then choose to bet on 1, 2, 3, 4, 6, 12, or 18 numbers—all of the standard roulette bets except the basket bet, which was not an option since Stakes and Odds games disregarded the 0 and 00 spaces—or could opt to pass on the wager.

The Stakes bets saw the casino defining the odds by specifying the number of numbers (1, 2, 3, 4, 6, 12, or 18 again) which the player would cover with his or her bet. If the player accepted the bet, he or she chose the amount to risk from the amounts listed above and selected which numbers on the layout to back.

At its heart, Game #9 was an examination of players' comfort with variance. Since the expected value of every bet in the game was 0, the variance of each wager was the only thing that distinguished one of these bets from another. A $1 bet on n numbers under these terms has variance

$$\sigma^2 = \left(\frac{36-n}{n}\right)^2 \cdot P(\text{Win}) + (-1)^2 \cdot P(\text{Lose})$$

$$= \left(\frac{36-n}{n}\right)^2 \cdot \left(\frac{n}{36}\right) + (1) \cdot \left(\frac{36-n}{36}\right)$$

$$= \frac{36^2 - 72n + n^2}{n^2} \cdot \frac{n}{36} + \frac{36-n}{36}$$

$$= \frac{36-n}{n},$$

precisely the numerical payoff amount on a $1 wager, though the units (dollars vs. dollars squared) are different. Betting an amount $A different from $1 multiplied the variance by A^2, and so the variance could be managed in two ways. Increasing the wager increased the variance, but betting on more numbers decreased the variance.

Example 3.7. Suppose the computer set the stakes of the wager at $5. A player comfortable with up to 50 dollars^2 of variance who would make a single-number bet of $1 (variance = 35) would only make a 12-number bet with $5 at risk, where the variance is 50 dollars^2. That same player might pass on a mandatory $25 bet, where the smallest variance, on an 18-number bet, is 625 dollars^2. ∎

A variance-averse player would stick to bets with a higher probability of winning and would risk less money if forced to cover fewer numbers, while a player with a high tolerance for variance would show that in a willingness to make riskier bets overall. It is unlikely that anyone playing Stakes and Odds had this kind of insight into their tolerance of variance, but Game #9 was designed to bring that information out through game play. Over the course of

14 wagers, many gamblers would display a consistent level of risk tolerance or aversion even if they were not aware of their personal preferences.

3.2   Mouse Roulette

> *You can't fix a mouse.*
>
> —Everett MacDonald, inventor of Mouse Roulette [56].

Crooked roulette wheels, some involving electromagnets that could be hidden in the table and used to control a roulette ball with a steel slug at its center, were not unheard of in underground casinos in the 1930s [89]. *Mouse roulette*, also known as *Cardette*, was a short-lived game that purported to remove all possibility of corruption by replacing the ball with a common house mouse.

The modified wheel was surrounded by 56 holes, each one leading to a glass trap. Each hole bore one of the 52 playing card symbols or one of several extra symbols. In play, the wheel was spun and the mouse dropped into the center. The winning symbol was determined by which glass trap the mouse entered. The wagers offered were similar to roulette bets, and are shown in Table 3.2.

TABLE 3.2: Mouse Roulette wagers [55]

Number of cards	Payoff odds
1	50 for 1
2	25 for 1
4	12 for 1
8	6 for 1
13 (Suit)	4 for 1
26	1–1
Even-money bets could be made on red, black, high, or low.	

Most wagers at American roulette carry a 5.26% HA due to the presence of the 2 zeroes on the wheel. This means that a single-number wager, for example, pays 35–1 on what is really a 37–1 shot. Since the payoff is 50 for 1, effectively 49 to 1, mouse roulette's single-card wager has an expected value of

$$E = (49) \cdot \frac{1}{56} + (-1) \cdot \frac{55}{56} = -\frac{6}{56} \approx -.1071,$$

so the HA is more than twice as high: 10.71%.

Mouse roulette debuted at Harolds Club in Reno, Nevada on May 15, 1936—and lasted one night. According to Raymond I. Smith, owner of Harolds Club, mouse roulette attracted a lot of curious spectators, but very few gamblers [61].

3.3 La Boule and French Bull

La Boule (or *Boule*) and *French Bull* are variations on roulette that use a rubber ball tossed into a wheel or tray, which is allowed to bounce around at random before coming to rest on a single winning digit from 1–9. As the names suggest, these games are of French origin, though French Bull's strongest foothold may have been in the casinos of southeast Asia [40, 82].

La Boule uses a wheel, which is much simpler than a roulette wheel, on which only the numbers 1 to 9 appear. A rubber ball is thrown into the spinning wheel and eventually settles into a hole at the center. Depending on the wheel, each number has two or four holes allocated to it. Figure 3.2 shows a La Boule wheel with 18 holes: two per number.

FIGURE 3.2: La Boule wheel [73]

The numbers 1, 3, 6, and 8 are black; 2, 4, 7, and 9 are red. The 5 is yellow and functions somewhat like the green zero in standard roulette. Table 3.3 shows the bets that are available at La Boule:

Notice that the payoff structure of La Boule is considerably less complex than standard roulette—only two different payoffs are available. The yellow number 5 is the source of the casino's edge on the even-money bets, as it is classified as neither odd, even, high, nor low. The even-money bets have a

TABLE 3.3: La Boule wagering options

Bet	Payoff	Description
Red/Black	1 to 1	Bet on red or black
Low (Manqué)	1 to 1	Bet on 1, 2, 3, 4
High (Passé)	1 to 1	Bet on 6, 7, 8, 9
Odd/Even	1 to 1	Bet on odd or even
Single number	7 to 1	Bet on any one number

common expectation given by

$$E = (1) \cdot \frac{4}{9} + (-1) \cdot \frac{5}{9} = -\frac{1}{9} \approx -\$.1111,$$

and the casino holds an 11.11% edge, over double the house advantage at American roulette. The same 11.11% edge accrues to the casino on single-number bets.

Recall this maxim from page 20:

> The easier a bet is to understand, the higher the house advantage.

This stands up to mathematical scrutiny, at least as a rough guideline. While roulette is about as easily understood as casino games get—pick some numbers and see if one of them shows up when a wheel is spun—La Boule makes even that game much simpler, and players who appreciate that simplicity pay for it in the form of a double-digit casino advantage.

French Bull eschews a wheel for a tray containing 67 numbered and indented circular pockets, into which a rubber ball is tossed and bounces around before settling into a circle and identifying a winning digit [40]. In this aspect, French Bull looks more like an actual carnival game, many of which involve tossing balls, than a traditional casino game. French Bull uses the digits from 0 to 9, which means that the 10 digits are not equally represented among the 67 pockets. Figure 3.3 shows how the 67 pockets are laid out.

Counting in Figure 3.3 shows that the digits 1–7 appear 7 times each, and 8, 9, and 0 have only 6 assigned pockets. This has some implications for wagering.

A single-number bet in French Bull pays 8–1, regardless of the number chosen. The expected value of a bet on one of the digits 1–7 is

$$E = (8) \cdot \frac{7}{67} + (-1) \cdot \frac{60}{67} = -\frac{4}{67} \approx -\$.0597,$$

and the corresponding expectation when choosing 0, 8, or 9 is

$$E = (8) \cdot \frac{6}{67} + (-1) \cdot \frac{61}{67} = -\frac{13}{67} \approx -\$.1940,$$

1		2		3		4		5		6		7		8		9		0
	1		2		3		4		5		6		7		8		9	
0		1		2		3		4		5		6		7		8		9
	0		1		2		3		4		5		6		7		8	
9		0		1		2		3		4		5		6		7		8
	9		0		1		2		3		4		5		6		7	
8		9		0		1		2		3		4		5		6		7

FIGURE 3.3: French Bull numbered tray.

which is over three times worse for the gambler. A small change in the frequency of the number on the board leads to a large relative difference in the house advantage.

French Bull offers three other bets, each on designated sets of 3 numbers, that pay 2–1. Of these, the bet on 1, 4, and 7 is better for the gambler than the bets on 2, 5, and 8 or on 3, 6, and 9, since the 1–4–7 bet has 21 winning pockets, while 2–5–8 and 3–6–9 have only 20 apiece. In all cases, a ball landing in a 0 pocket wins all three-number bets for the house. If X is the number of winning pockets, then these three bets have an expected value of

$$E = (2) \cdot \frac{X}{67} + (-1) \cdot \frac{67 - X}{67} = \frac{3X - 67}{67},$$

which is negative, since $3X < 67$ no matter which bet is chosen. The 1–4–7 bet has the lower house edge, at 5.97%, while the casino's advantage on the other two three-number bets is nearly double that: 10.45%.

Example 3.8. If the French Bull board were expanded to equalize the digit frequencies by including one additional pocket for each of the numbers 8, 9, and 0, what edge would the casino retain on the two bets described above?

The single-number bets would each have an expectation of

$$E = (8) \cdot \frac{7}{70} + (-1) \cdot \frac{63}{70} = -\frac{7}{70} = -\$.10,$$

which settles the HA at a constant 10% on every number. For the three-number bets, the 0 would still be a losing number for players, whose expectation would be

$$E = (2) \cdot \frac{21}{70} + (-1) \cdot \frac{49}{70} = -\$.10,$$

conceding the same 10% edge to the house. ∎

3.4   Big Six Wheel

The *Big Six* wheel is a large wheel (usually six feet or more in diameter; the wheel at Vegas World measured 30 feet across) divided into 54 sectors. Each sector bears some kind of symbol, originally a piece of currency or casino logo. These wheels are frequently placed near the perimeter of the casino floor in full view of an entrance, as if to beckon a prospective gambler with the lure of a fun and simple game.

Fun? If you're winning, probably. Every game is fun when you're winning. Simple? Yes. Players bet on one of the wheel's symbols. The wheel is spun, and if the symbol at the top of the wheel, indicated by an arrow or a flexible strip called a *clapper*, matches the one the player bets on, the bet wins.

The Big Six wheel draws its name from the number of different payoffs that are possible when it's spun. A standard wheel is divided into 54 sectors, and the following symbols appear in the quantities shown in Table 3.4.

TABLE 3.4: Standard Big Six wheel symbols

Symbol	Count
$1	24
$2	15
$5	7
$10	4
$20	2
Logo A	1
Logo B	1

At the Golden Nugget Casino in downtown Las Vegas, the two logos on the wheel depict the H_2O Bar and the Hideout, two establishments located near the resort's swimming pool. On the Strip at Mandalay Bay, the logo slots bear a tropical flower and are colored red and blue to distinguish them. Currency spaces pay their face value on a win; logo spaces pay off at 40–1 in most jurisdictions. In New Jersey, state law requires that logo spaces pay 45–1.

Calculating probabilities on the Big Six wheel is a simple matter of counting the sectors containing the symbol on which you have bet and dividing by 54. Accordingly, we have the probabilities shown in Table 3.5. The best bet on the Big Six wheel, where "best" is defined as "lowest HA," is on the $1 spot. "Lowest," however, need not mean "low," as this bet carries a HA of 11.11%. Other bets have HAs ranging from 16.67% to 24.07%.

The symbol distribution and corresponding payoff values are certainly open for modification as a casino may wish. In particular, there is no requirement that payoffs correspond to currency denominations, or that currency sym-

TABLE 3.5: Standard Big Six wheel probabilities

Symbol	Probability
$1	.4444
$2	.2778
$5	.1296
$10	.0741
$20	.0370
Logo A	.0185
Logo B	.0185

bols be used at all. A wheel at the Soaring Eagle Casino in Mount Pleasant, Michigan used the symbols shown in Table 3.6. Numbered spaces pay their numerical value; the American flag and Joker spaces pay 50–1.

TABLE 3.6: Big Six wheel symbols: Soaring Eagle Casino

Symbol	Count
1	26
3	13
6	7
12	4
25	2
Flag	1
Joker	1

This wheel is somewhat better for players than the standard Big Six wheel; a bet on the $1, $3, $12, or $25 spots faces a HA of only 3.70%; the house edges on the other bets are all less than the 11.11% HA of the best standard Big Six bet. These lower HAs are possible because Interblock, the game's manufacturer, has configured it as an all-electronic game which requires no human dealer. The cost of operating the game is decreased considerably, and some of the savings can then be passed on to patrons. A second advantage to a better pay table is that patrons might win more frequently or win larger amounts. Passersby might then be drawn in by the attendant excitement, and some might be encouraged to play a game that has a reputation for being unfriendly to players.

Another electronic Big Six wheel is marketed by Aruze Gaming as the *Lucky Big Wheel*, with slightly different payoff values and frequencies. This wheel is marketed with several different symbol distributions; one version, installed at the Stratosphere Casino in Las Vegas in 2017, is shown with the HA of each bet in Table 3.7. As at the Soaring Eagle, the flag and joker symbols pay off at 50–1, which makes them the best bets on this wheel.

TABLE 3.7: Lucky Big Wheel symbols and house advantages

Symbol	Count	HA
1	23	14.81%
2	15	16.67%
5	8	11.11%
11	4	11.11%
24	2	7.41%
Flag	1	5.56%
Joker	1	5.56%

Dream Catcher, from Evolution Gaming, is a Big Six wheel variation in a live online casino, where a dealer spins the wheel and interacts with online players. The wheel includes two bonus multiplier sectors, which replace the special logos. The breakdown of its prizes is shown in Table 3.8.

TABLE 3.8: Dream Catcher wheel symbols [49]

Symbol	Count
1	23
2	15
5	7
10	4
20	2
2×	1
7×	1

If a wheel spin turns up either one of the multipliers, all bets are held and the wheel is respun until a number appears—subsequent spins landing on multipliers are disregarded. Once a winning number is chosen, all payoffs are multiplied by that multiplier, so a $1 bet on the 20 spot could pay off $140 if the 7× and 20 sectors are spun in succession. The probability of this payoff is

$$\frac{1}{54} \cdot \frac{2}{54} = \frac{1}{1458}.$$

Unlikely though this may be, the presence of the multipliers nonetheless decreases the large HA on the wheel.

Example 3.9. Recall that the standard Big Six configuration has an 11.11% HA when betting on the 1. Since successive spins of the Dream Catcher wheel are independent, the probability of winning $7 when betting on the 1 spot is

$$\frac{1}{54} \cdot \frac{23}{52} = \frac{23}{2808}.$$

Here, the denominator in the second fraction, which represents the chance of spinning a 1 after the 7× sector has come up, is 52 rather than 54 since the multiplier spaces are disregarded on that spin. The complete probability distribution for this bet is shown in Table 3.9.

TABLE 3.9: Dream Catcher probability distribution: $1 bet on the 1 spot

Outcome	Probability
$1	$\dfrac{23}{54}$
$2	$\dfrac{23}{2808}$
$7	$\dfrac{23}{2808}$
–$1	$\dfrac{29}{52}$

The HA of this bet is then only 5.80%, about half of the HA on a standard wheel. ∎

The *Federal Wheel* is a version of the Big Six wheel that is approved for use in Australian casinos. At the Wrest Point Casino in Tasmania, the wheel is divided into 50 spaces rather than the 54 of the Big Six wheel, and each space is labeled with the name of an Australian state or territory as in Table 3.10 [39].

TABLE 3.10: Federal Wheel space distribution [39]

State/Territory	Spaces	Payoff Odds
Tasmania	24	1–1
New South Wales	8	5–1
Victoria	8	5–1
Queensland	4	11–1
South Australia	2	23–1
Western Australia	2	23–1
Australian Capital Territory	1	47–1
Northern Territory	1	47–1

Every wager on this wheel has the same house advantage. We can find it in two steps, beginning by finding an algebraic expression for the payoff x (to 1) in terms of n, the number of spaces on which a state or territory appears. Careful counting, inspired in part by the payoff formula for most roulette wagers, gives

$$x(n) = \frac{48 - n}{n}.$$

We now use this expression to find the expectation and house advantage for a $1 wager.

$$E = \left(\frac{48 - n}{n}\right) \cdot \frac{n}{50} + (-1) \cdot \frac{50 - n}{50} = -\frac{2}{50} = -\$.04,$$

so the HA is 4%.

This is markedly better for the player than the 11.11–24.07% HA at the Big Six wheel. In the words of gambling writer Syd Helprin, though, "This edge may not be low enough to cause gamers to take off hurriedly for Australia" [39].

Mystery Card Bonanza

The Big Six wheel, in any of its configurations, often sits idle on the casino floor, in part, perhaps, because serious gamblers are well aware that its house advantage is far higher than other table games. With *Mystery Card Bonanza*, INAG International promoted a version of the Big Six wheel with a lower house edge on certain wagers.

This wheel has been determined to be a card-shuffling device and not a "Wheel of Fortune," which are outlawed in some gaming jurisdictions. As such, the inventors are confident that it satisfies the legal requirements for most states. The currency symbols on a standard Big Six wheel are replaced by cards from a custom 54-card deck including aces, 2s, 5s, 10s, kings, and two logo cards [62]. The wheel is spun to select a card, which determines the payoffs; players may bet on a rank, a suit, a color, or either logo.

Example 3.10. For bets on individual card ranks, the payoffs mimic the odds on a standard Big Six wheel, with the king standing in for the $20 bill and all other bills replaced by their numerical value. Table 3.11 shows the game information.

TABLE 3.11: Mystery Card Bonanza pay table

Card	Count	Payoff
Ace	24	1–1
2	15	2–1
5	7	5–1
10	4	10–1
King	2	20–1
Logo A	1	40–1
Logo B	1	40–1

These bets have the same high house advantages as an ordinary Big Six wheel. However, since the four suits are equally represented among the 52 playing cards and a suit bet pays off at 3–1, the expected value of suit bets is

$$E = (3) \cdot \left(\frac{13}{54}\right) + (-10) \cdot \left(\frac{41}{54}\right) = -\frac{2}{54} \approx -\$.0370,$$

for a considerably lower HA of 3.70%. ∎

An even-money bet on red or black is similarly less unfavorable to the player; its expected value is also –$.0370 per dollar wagered.

The Mystery Card Bonanza layout contains space for 8 bettors, and can be configured to accommodate an optional electronic Hotspot Bonus bet [62]. This simple wager, which has no connection to the play of the game, uses a random number generator to select and light up a winning player position; that player is paid off at 6–1 *if* he or she has made this bet. Its simplicity leads us to suspect that the HA here is high, and the casino does indeed enjoy a sizable edge: 12.5%.

Pokette

The 54 sectors on a standard Big Six wheel correspond nicely to a 52-card deck with 2 jokers. *Pokette*'s designers took this correspondence and turned it into a new table game in 1992. The game was also approved for play on a special 54-pocket roulette wheel [10, 115]. Pokette, also known as *Card-o-lette*, was a poker-roulette hybrid introduced at the TropWorld Casino in Atlantic City, New Jersey (now the Atlantic City Tropicana).

A Pokette wheel bore the images of the 52 cards in a standard deck and devoted 2 sectors to jokers. The Pokette layout was organized with 4 columns of card ranks, corresponding to suits. Spaces for the two jokers were at the head of the layout, just as the zeros are placed on a roulette layout. Several wagers were available, with varying numbers of cards and spins involved.

At the start of a new Pokette hand, players were invited to place **1-spin bets**, which worked very much like roulette bets. These were bets on one or more cards, and paid off on the next spin of the wheel. They are listed with their payoff odds in Table 3.12. Unless $n = 8$, a wager on n cards pays off at the easily remembered odds of $\dfrac{52 - n}{n}$ to 1.

TABLE 3.12: Pokette: 1-spin wagers [117]

Wager	Payoff Odds
Single card	51–1
Two cards	25–1
Four cards	12–1
Eight cards (2 ranks)	5–1
Three ranks (13 cards)	3–1
13 cards (Suit bet)	3–1
Red/black (26 cards)	1–1

Four-card bets were available on a single rank or on 4 cards forming a square on the layout. Three-rank bets could be made on AKQ, J109, 876, and 543; each block of cards also includes a designated wild deuce, bringing the total number of cards to 13.

An extra 1-spin Pokette bet was the *Jacer* bet, which was a 4-card bet covering the 2 jokers and the aces of clubs and diamonds [85]. As a bet on 4 cards, it paid 12–1. This bet is somewhat analogous to American roulette's 5-number basket bet, as it involves the jokers and the two cards nearest them on the Pokette layout.

Following the first spin, all bets were resolved. Players were then invited to make new single-spin wagers or, if the chosen card was not a joker, could make the sole **2-spin bet**: an 11–1 wager that the second spin would pair the first card. Since the successive spins were independent, it was possible to repeat cards, and thus a pair like $A\heartsuit\,A\heartsuit$ could occur. A pair of jokers would not win this bet, since no Pair bets were accepted on an initial joker.

After the second card was chosen, and with no jokers yet appearing, a new array of **3-spin bets** was available in addition to the previously active wagers. These were wagers on the 3-card poker hand formed by 3 consecutive spins. Actual playing cards were used to track successive spins, with cards inserted into a tableside display rack following spins 1 and 2. Table 3.13 shows the 3-card wagering options; only those which were still possible after 2 cards were determined were available for player action [85].

TABLE 3.13: Pokette: 3-spin wagers [117]

Wager	Payoff Odds
Inside straight flush	49–1
Open straight flush	24–1
Flush	3–1
Inside straight	11–1
Open straight	5–1
3 of a kind	11–1
Pair in 3 cards	5–1

For straights and straight flushes, "inside" wagers would be active if the first two cards were one apart in rank, as with the $K\Diamond$ and $J\Diamond$ or $8\spadesuit$ and $6\Diamond$. "Open" straights and straight flushes were betting options if the first two cards were consecutive, for example, the $J\clubsuit$ and $10\spadesuit$. Open hands have twice as many ways to win as inside hands, and their payoff odds are, accordingly, lower.

Example 3.11. If the first two cards were the $7\heartsuit$ and $5\Diamond$, the only 3-spin bets accepted by the casino would be the inside straight and pair in 3 cards wagers, since no other bets had a chance of winning. ∎

In poker played with cards, jokers are frequently used as wild cards, which add to the player's edge. Jokers are *not* wild in Pokette. The jokers may have been an attractive game element that drew players in, and an unwelcome surprise during gameplay that drove them away. In Pokette, the jokers are the source of the house advantage. Consider the 2-spin Pair bet. If jokers are wild, a joker on the first card would mean that all Pair bets on the next spin would win, which would ruin the casino as players rushed to wager a fortune on Pair and win 11 times their stake with no risk at all.

With the jokers losing for the players, the probability of winning the 2-card Pair bet is $\frac{4}{54} \approx .0741$. The corresponding expectation is

$$E = (11) \cdot \frac{4}{54} + (-1) \cdot \frac{50}{54} = -\frac{6}{54} \approx -.1111,$$

giving the casino an 11.11% advantage.

It should be noted that a player considering the Pair bet would be better off making a 1-card bet on the rank of the card just spun. This 4-card bet pays off at 12–1 rather than 11–1, and the corresponding HA is only 3.70%—one-third that of the Pair bet.

Example 3.12. A player would make the 3-spin Pair bet after the first two spins turn up different non-joker cards. This means trading better payoff odds relative to the 2-card Pair bet for a slightly increased chance to draw a pair. Which Pair bet is better for the player?

The HA of the 2-spin Pair bet was shown above to be 11.11%. If the first two cards are neither jokers nor of the same rank, this bet has

$$P(\text{Win}) = \frac{8}{54} \approx .1481$$

and expected value

$$E = (5) \cdot \frac{8}{54} + (-1) \cdot \frac{46}{54} = -\frac{6}{54} \approx -\$.1111.$$

The HA is the same: 11.11%. ∎

Once again, though, it's better for the gambler simply to make a 1-card bet on the rank of the first 2 cards. The HA then drops to 3.70%, just as it did with the 2-card Pair bet.

Pokette was discontinued in Atlantic City in 1993, only a year after its introduction, though the game remains approved for play in New Jersey and the rules are regulated by the New Jersey Division of Gaming Enforcement [85, 116].

Minnesota Tri-Wheel

Minnesota state law is quite restrictive in games of chance permitted at charitable fundraisers and for casual play in bars, which carries over into games that may legally be offered at the state's Indian casinos. While this list does not include craps, roulette, or most card games other than blackjack, a class of games called *paddlewheel* games, which are similar to the Big Six wheel, is permitted. Authorized in 1987, the *Minnesota Tri-Wheel* is a paddlewheel game that involves 3 concentric rings of numbers, allowing players to select wagers that fit their comfort with uncertainty [29]. The innermost wheel is colored orange and is divided into 10 sectors numbered 1–10; the middle wheel is yellow and its sectors are numbered 1–20; and the outside wheel is blue and numbered 1–40. When the wheels are spun, they rotate independently, and the clapper designates 3 winning numbers, one per wheel, when they stop. This game is available in a number of bars and restaurants in Minnesota.

The layout for bets on the blue wheel is shown here.

A	1	2	3	4	5	6	7	8	9	10	A
B	11	12	13	14	15	16	17	18	19	20	B
C	21	22	23	24	25	26	27	28	29	30	C
D	31	32	33	34	35	36	37	28	39	40	D

Rows labeled A–D identify sets of numbers on which players may place Line bets covering all the numbers in one row.

The yellow layout looks like this, with Line bets available on five groups of 4 numbers each.

E	1	2	3	4
F	5	6	7	8
G	9	10	11	12
H	13	14	15	16
I	17	18	19	20

The 10-number orange wheel permits no line bets, and so its layout is a single line of 10 boxes, one for each number from 1–10.

Tri-Wheel wagering is similar to roulette: players choose one or more wheels and may then bet on single numbers, on lines within the yellow and blue layouts, or on Odd or Even within the blue wheel. Unlike in roulette, though, players may not place bets on the interior of the layout covering more than one number. The pay table is shown in Table 3.14; note that all payoffs are "for 1."

A quick look at Table 3.14 shows that the Odd and Even bets appear to have no house advantage, since half of the numbers on the blue wheel are even and half are odd. The house retains an edge by placing stickers on 6 blue numbers, designating them as "house" numbers on which Odd or Even bets all lose. Typically, these numbers are roughly evenly spaced around the wheel; 3 are odd and 3 even. The stickers only apply to Odd and Even bets; bets made on a single stickered number or on a line including one pay off normally.

TABLE 3.14: Minnesota Tri-Wheel pay table [79]

Wager	Payoff Odds
Single-number: Blue	35 for 1
Single-number: Yellow	17 for 1
Single-number: Orange	8 for 1
Line bet: Blue (10 numbers)	3 for 1
Line bet: Yellow (4 numbers)	4 for 1
Odd/Even (Blue wheel only)	2 for 1

With this modification in place the expected value of an Odd or Even bet paying 2 for 1 is

$$E = (2) \cdot \frac{17}{40} - 1 = -\frac{6}{40} = -\$.15.$$

The HA is 15%; promotional material for the game notes that this bet returns 85%, which is equivalent but, by focusing on the win percentage, gives a different picture of the true nature of this bet. Gaming Studio, Inc., designer of the wheel, asserts a difference between "gaming" and "gambling," and notes on its Web site that players should "Game for fun—Gamble for profit." Furthermore, the company urges: "Don't confuse gaming for gambling—our games will not pay your rent" [29]. At a 15% house edge, this is certainly a true statement.

High HAs, of course, are common when a game is intended as a charitable fundraiser, as with state lottery games that typically have a house advantage around 50%. Minnesota Tri-Wheel continues its high house edges in its single-number bets.

Blue: $E = (35) \cdot \dfrac{1}{40} - 1 = -\$.125.$ HA = 12.5%.

Yellow: $E = (17) \cdot \dfrac{1}{20} - 1 = -\$.15.$ HA = 15%.

Orange: $E = (8) \cdot \dfrac{1}{10} - 1 = -\$.20.$ HA = 20%.

Example 3.13. How does the HA on the Odd and Even bets vary with the number of stickered numbers on the blue wheel?

Let $2x, x \geq 0$, be the number of stickers applied to the wheel. For symmetry between the bets, x of the stickers should be on odd numbers and x on even numbers. The expected value of a \$1 bet as a function of x is then

$$E(x) = (2) \cdot \frac{20 - x}{40} - 1 = -\frac{x}{20}.$$

If 2 stickers are used, the HA is 5%—on the edge of a reasonable advantage for a casino game. Four stickers ($x = 2$) gives a 10% HA, and the HA increases by 5% for every additional pair of stickers. ∎

3.5   Rouleno

As the name might suggest, *Rouleno* was described as a combination of roulette and keno. All three games have outcomes determined by balls of some sort. Rouleno appeared at the Max Casino in Las Vegas in October 2016. The game made its debut shortly after the casino's home resort, the Westin Las Vegas, announced that the casino would close, but the closure was delayed about a year and the casino remained open until mid-2017. The Westin hotel remains open, with the casino space replaced by a restaurant.

Rouleno was a highly unconventional carnival game, both in terms of the equipment it used and the mathematics necessary to analyze it. The game was played with six standard pool balls numbered 1–6, which were mixed in a wooden box and drawn successively. Players were able to wager on how many of the balls were drawn in the "correct" ordered position, as when the 1 ball was drawn first, the 2 ball second, and so on. The game did not require pool balls; it could easily have been played with 6 playing cards. However, acceptably shuffling a 6-card deck would have been difficult, leaving a card-based version of Rouleno susceptible to shuffle tracking.

Example 3.14. If the balls are drawn in the order

3	6	2	1	5	4

then one ball, the 5, appears in the correct position.   ■

Since Rouleno is based on the number of balls drawn into the right position, order matters when computing the associated probabilities. There are

$$_6P_6 = 6! = 720$$

ways to arrange the 6 balls with order considered. To compute the probability of k balls landing in the correct spot, we are interested in counting both the number of ways to draw some balls in the right places and the number of ways to draw the other balls in the wrong places. If we denote the number of matches by the random variable X, this second factor ensures that we are computing $P(X = k)$ rather than $P(X \geqslant k)$.

Counting the number of ways to draw some balls into incorrect places calls for a type of permutation called a *derangement*.

Definition 3.1. A *derangement* of a finite set is a permutation of the elements of the set which has no fixed points: every element of the set is moved from its original position. The number of derangements of a set with n elements is denoted D_n. By definition, $D_0 = 1$.

Forming a derangement of a given set is known as *deranging* the set.

It is convenient to denote the set by $\{1, 2, 3, \ldots, n\}$ so it is clearer when an element has been moved.

Example 3.15. The Rouleno draw in Example 3.14 is not a derangement, since the 5 is in the correct place. If the balls are drawn into the order

4	3	5	2	6	1

we have a derangement. ∎

Since every derangement is a permutation of the entire set of n elements, yet not every permutation is a derangement, we know that $D_n < {_n}P_n = n!$. The definition establishes that $D_0 = 1$. $D_1 = 0$, since there is no way to derange the set $\{1\}$. Since the only way to derange the set $\{1, 2\}$ is to swap the elements to form $\{2, 1\}$, we have $D_2 = 1$.

D_3 can be determined by careful reasoning. We know that the first element in a derangement of $\{1, 2, 3\}$ cannot be 1. Choosing which of 2 and 3 goes into the first slot determines the entire derangement. If the derangement begins with 2, then 3 must be in the second slot since it cannot be third, leaving 1 for the third slot. We see that the only derangement of $\{1, 2, 3\}$ beginning with 2 is $\{2, 3, 1\}$. Similarly, if the derangement starts out with 3, then since 2 cannot go second, the only valid derangement is $\{3, 1, 2\}$. Consequently, $D_3 = 2$.

As with permutations and combinations, we are typically not as interested in generating a list of all derangements of a set of n numbers as in counting how many derangements are possible for a given set. A simple formula for D_n when $n \geqslant 1$ is given by

$$D_n = \left\lfloor \frac{n!}{e} + \frac{1}{2} \right\rfloor,$$

where $e \approx 2.71828...$ is the base of the natural logarithm and $\lfloor x \rfloor$ denotes the *floor function*: the greatest integer less than or equal to x [7]. Table 3.15 gives values of D_n which are pertinent to Rouleno.

TABLE 3.15: Derangements

n	0	1	2	3	4	5	6
D_n	1	0	1	2	9	44	265

Suppose that k numbers are drawn correctly. This may be done in $\binom{6}{k}$ ways: though we are solving a problem where order matters, we have no choice as to where the correct balls go once they've been identified. The remaining $6 - k$ slots are to be filled with the remaining $6 - k$ balls, with the provision that the balls are deranged into new positions. This may be done in D_{6-k} ways. Accordingly,

$$P(k \text{ matches}) = \frac{\binom{6}{k} \cdot D_{6-k}}{{_6}P_6} = \frac{\binom{6}{k} \cdot D_{6-k}}{720}.$$

This probability distribution function for Rouleno is compiled in Table 3.16.

TABLE 3.16: Probability distribution function for Rouleno

k	$P(k$ matches$)$
0	.3681
1	.3667
2	.1875
3	.0556
4	.0208
5	0
6	.0014

Table 3.16 shows $P(5) = 0$, so that drawing exactly 5 numbers into the right places is impossible. This is because if 5 balls are chosen and placed correctly, there is only one slot left for the sixth ball, and it is the correct place for that ball. Mathematically, this result follows from the fact that $D_1 = 0$.

The expected number of correctly drawn balls is

$$\sum_{k=0}^{6} k \cdot P(k \text{ matches}) = 1,$$

While this is nearly tied with 0 matches for the most likely result, it occurs less than 40% of the time.

Example 3.16. If the number of Rouleno balls is not limited to 6, but is allowed to vary and denoted by n, we have the more general PDF

$$P(k \text{ matches}) = \frac{\binom{n}{k} \cdot D_{n-k}}{n!}.$$

This expression may be expanded, giving

$$P(k \text{ matches}) \approx \frac{\dfrac{n!}{(n-k)! \cdot k!} \cdot \dfrac{(n-k)!}{e}}{n!}.$$

As the number of balls n increases, the approximation

$$D_{n-k} \approx \frac{(n-k)!}{e}$$

improves in accuracy, and we find that $P(k$ matches$)$ approaches $\dfrac{1}{k! \cdot e}$.

If $k = 0$, we have

$$P(0 \text{ matches}) \approx \frac{1}{e} \approx .3679,$$

a result which is independent of n. Since $0! = 1! = 1$, it follows that $P(1$ match$)$ also approaches $1/e$. Moreover, the expected number of matches when n balls are used remains 1, as it was with $n = 6$. ■

We turn now to the available Rouleno wagers. The simplest bet is an individual number bet, which pays off at 5–1 if the selected number is drawn in the correct spot. As written, this is a fair bet; the casino extracts an edge by designating three losing sequences. All individual number bets lose if any of the following three sequences are drawn:

$$1\ 2\ 3\ 4\ 6\ 5$$
$$1\ 2\ 4\ 3\ 5\ 6$$
$$2\ 1\ 3\ 4\ 5\ 6.$$

Only two of these sequences turn a given winning individual number bet into a loser: for example, a bet on 3 is already lost if the sequence 1 2 4 3 5 6 is drawn.

Since there are $_5P_5 = 120$ ways to draw the other 5 numbers into 5 spots once the chosen number is drawn into its correct position, the probability of winning an individual number bet is then

$$\frac{_5P_5 - 2}{_6P_6} = \frac{118}{720}.$$

The expected value is

$$E = (5) \cdot \frac{118}{720} + (-1) \cdot \frac{602}{720} = -\frac{12}{720} = -\$\frac{1}{60} \approx -\$.0167,$$

which results in a very low HA of 1.67%, comparable to the Pass and Don't Pass bets at craps and better than the 2.70% edge the casino holds at European roulette.

An inattentive dealer who overlooks the three automatic losing sequences, one of which should occur about once in 240 draws, and pays off on all correctly drawn individual numbers is dealing a fair game.

A player wishing to insure an individual number bet against one of the losing sequences can do so by placing an *Any 4* bet: a bet that exactly 4 numbers will be drawn correctly, which pays off at 40–1. There are no house sequences excluded from this bet; any of the draws that put exactly 4 numbers in the right spots score a win. Following the formula above, there are

$$\binom{6}{4} \cdot D_2 = 15 \cdot 1 = 15$$

of these.

This insurance is costly; the expected value of such a wager is

$$E = (40) \cdot \frac{15}{720} + (-1) \cdot \frac{705}{720} = -\frac{105}{720} \approx -\$.1458,$$

giving the house a 14.58% edge.

The two outcomes "Match 0" and "Match 1," which have approximately equal probabilities, account together for just over 73% of all outcomes. Players may choose the *Match 0* or *Match 1* bets, which pay off at 3–2 on the selected number of matches *or* if exactly 4 balls match. This provides an additional 15 winning draws. The exact probability of 0 matches is

$$\frac{D_6}{{}_6P_6} = \frac{265}{720} \approx .3681,$$

and the probability of 1 match is

$$\frac{\binom{6}{1} \cdot D_5}{{}_6P_6} = \frac{264}{720} \approx .3667.$$

If the Match 4 outcomes are not included as winners, the expected value of $2 wagered on either bet is approximately –16¢ and the HA is over 8%. Adding in 15 more winning outcomes raises the expectation to about –6¢, which makes the HA close to 3% on either bet.

There is only one way for all 6 balls to be drawn in order. This is called a "Perfect" drawing, or a *rouleno*. The *Perfect* bet also pays off if the balls are drawn in the reverse order 6 5 4 3 2 1: the "Perfect Reverse" drawing. This gives 2 winning possibilities for a bet that pays 300 for 1, or 299–1. Carrying a 16.67% HA, this is the worst bet on the board.

The *Any 2* and *Any 3* bets are straightforward wagers with no additional winning combinations added or losing combinations excluded. They are summarized together with the bets we have so far examined in Table 3.17.

TABLE 3.17: Summary of Rouleno bets

Wager	P(Win)	Payoff	HA
Individual number	.1639	5–1	1.67%
Zero + Any 4	.3889	3–2	2.78%
One + Any 4	.3875	3–2	3.13%
Any Two	.1875	4–1	6.25%
Any Three	.0556	15–1	11.11%
Any Four	.0208	40–1	14.58%
Perfect*	.0014	299–1	16.67%
*—Includes Perfect Reverse			

Rouleno did not outlast the Max Casino; while it remains on the Nevada Gaming Control Board's list of approved casino games [64], it is not currently available for play in any Nevada casino.

3.6 Exercises

Solutions begin on page 286.

Stakes and Odds

1. Stakes and Odds Game #5 was an experiment that, like Game #9, tested subjects' comfort with variance. Players were presented with a list of 11 bets and chose one on which to bet. The player's choice determined the number of "win groups," or rows of 3 numbers on the roulette layout that won for the player, so a choice of n numbers, where $1 \leqslant n \leqslant 11$, generated $3n$ winning numbers. One set of wagers is shown with its payoffs and risks in Table 3.18.

TABLE 3.18: Stakes and Odds: Game 5

Win Groups, n	Amount won	Amount lost
1	66.00	6.00
2	30.00	6.00
3	18.00	6.00
4	12.00	6.00
5	8.40	6.00
6	6.00	6.00
7	4.29	6.00
8	3.00	6.00
9	2.00	6.00
10	1.20	6.00
11	0.55	6.00

a. Let n be the number of win groups chosen by the player. Derive an expression for the winning payoff as a function of n.

b. Use the expression derived in part a to show that, to the nearest penny, each bet in Table 3.18 is fair.

c. Let $X(n)$ be a random variable measuring the outcome of one of these bets. Calculate the variance of $X(n)$ as a function of n. Notice that, unlike the expectation, the variance is not constant.

d. How does the standard deviation of $X(n)$ vary with n?

2. An additional feature of game #5 was the player's choice among large, medium, or small bets. The payoffs for one set of small bets are those shown in Table 3.18. Medium bets paid 3 times as much for a win and charged 3 times as much on a loss, while large bets multiplied both payoffs and losses by 5. What effect does the bet multiplier have on the variance and standard deviation of $X(n)$ as defined above?

3. Table 3.18 was also used in Stakes and Odds Game #10, as one of 27 similar lists of fair bets. Some lists followed Table 3.18 by varying the potential win and fixing the loss, while others fixed the amount to be won and varied what a player could lose. Players were encouraged to pick the bet that they liked best from each list as it was presented, and some lists allowed a player to increase his or her bet beyond the amount initially listed as at risk.

What kind of attitude toward risk could be measured by a player's decisions in Game #10?

Mouse Roulette

4. Suppose that mouse roulette wagers paid off at x to 1, rather than x for 1. Find the HA for the various bets listed in Table 3.2. Which bet has the best player edge?

5. European roulette guarantees the same HA on every wager by setting the payoff odds on an n-number wager at $\dfrac{36 - n}{n}$ to 1. If the mouse roulette payoffs are set at $\dfrac{52 - n}{n}$ to 1 for an n-number bet, find the common HA.

6. Repeat Exercise 5 with the same payoff odds under the assumption that the wheel had only 54 spaces: one for each playing card and two jokers.

Federal Wheel

7. An alternate Federal Wheel arrangement which includes one wild space and one semi-wild space is described in Table 3.19 [84].

If the wheel stops on "Joker," all bets pay off at 3–1, regardless of the payoff when their state or territory is spun.

If the wheel stops on "All States Win," then all bets on individual states win, except for bets on the Australian Capital Territory or Northern Territory, which push.

In contrast to the wheel on page 65, different states and territories sometimes offer different house advantages.

a. Confirm that wagers on South Australia and Western Australia have the same HA.

b. Which Australian state or territory offers the lowest house advantage on this wheel?

c. Which Australian state or territory offers the highest house advantage on this wheel?

TABLE 3.19: Federal Wheel space distribution [84]

State/Territory	Spaces	Payoff Odds
Tasmania	19	1–1
Victoria	12	2–1
New South Wales	5	6–1
Queensland	5	6–1
South Australia	3	13–1
Western Australia	2	20–1
Australian Capital Territory	1	41–1
Northern Territory	1	41–1
Joker	1	All bets pay 3–1.
All States Win	1	State bets win. Territory bets push.

Minnesota Tri-Wheel

8. Show that the Line bet on the blue wheel is the worst bet for a Minnesota Tri-Wheel player.

Pig Wheel

The Minnesota Tri-Wheel spun off a descendant, the *Pig Wheel*, which is approved for use in North Dakota [80]. Pig Wheel is a variation on the Big Six wheel that is less complicated than the Minnesota Tri-Wheel, consisting of a single 45-sector wheel. Forty spaces are numbered, from 1–40, and 5 spaces bear pictures of pigs: 2 blue (named "Joe" and "Bob"), 2 pink ("Roxy" and "Sue"), and 1 white ("Ken"). The pigs assure the house an advantage; they function somewhat like the stickered numbers in Minnesota Tri-Wheel without the player disappointment that might ensue when a gambler has bet on Odd and the number on the blue wheel is an odd number bearing a sticker. (Paddlewheels in Minnesota are restricted to numbers in labeling wheel sectors, so Pig Wheel is not approved for use in that state [80].)

Pig Wheel allows for slightly more wagers than Minnesota Tri-Wheel; the layout for bets on numbers is shown below.

A	1	2	3	4	5	6	7	8	9	10	A
B	11	12	13	14	15	16	17	18	19	20	**B**
C	21	22	23	24	25	26	27	28	29	30	**C**
D	31	32	33	34	35	36	37	28	39	40	**D**
	E	**F**	**G**	**H**	**I**	**J**	**K**	**L**	**M**	**N**	

Additional spaces adjacent to this grid allow for bets on pigs by color.

Pig Wheel offers the following wagers [80]:

- Single-number bets.
- Two-number bets, as are available in roulette on adjacent numbers, either horizontally or vertically.
- Four-number bets on blocks of 4 numbers on the layout or on the columns labeled E through N.
- Line bets on rows of 10 numbers labeled A–D.
- Odd and Even bets, which lose if a pig is spun.
- Pig bets, on the three different colors of pig. Pink and blue pig bets cover 2 spaces; the white pig bet has only 1 winning space.

9. The major bets at Pig Wheel are the single-number, Line, Pig, and Odd/Even bets. Given that these bets all have a house advantage of 11.11% ($\frac{1}{9}$), find the payoff odds (for 1) for each bet [80].

10. Payoff odds on a 2-number Pig Wheel bet are under operator control, and so the house advantage may vary among locations. Find the whole-number payoff odds for this bet so that the HA is as close to 11.11% as possible.

11. An "Any Pig" bet pays off if any of the 5 pig sectors is spun. If the HA is intended to be as close to 11.11% as possible, what should the payoff be?

Pokette

12. If the first two spins of a Pokette wheel turn up $Q\Diamond\ J\Diamond$, players would have the option to place either a Flush bet or an Open Straight bet. Which wager has the lower casino edge?

13. Suppose that the first two spins of a Pokette wheel turn up the same card, for example $5\Diamond\ 5\Diamond$. Which wager has the better player expectation: Three of a Kind or Flush? How do these HAs compare to that of a 1-card bet on the 5 rank?

Rouleno

14. Rouleno also offers a combination bet which collects several matches into one winning outcome. The bet pays off at 10–1 if 3 or more balls are drawn into the correct slots, or if a Reverse Perfect draw occurs. Find the probability of winning this bet.

15. Consider an extension of Rouleno where the remaining solid-color pool balls, the 7 and 8 balls, are added to the draw.

a. Calculate D_7 and D_8, which will be necessary in computing probabilities.

b. Generalize the list of barred permutations that turn a single-number bet into a loser by defining four losing permutations that will be excluded in this enhanced game.

c. Calculate the probability of winning a single-number bet with four permutations excluded, and determine payoff odds that give this bet a house edge between 0 and 5%.

d. Tabulate the PDF for the number X of successful matches.

e. Find the probability of winning a hypothetical "Zero + Any Six" wager.

f. The analog of the combination bet described in Exercise 14 would pay off if 4 or more balls are drawn into the correct slots, or if an 8-ball Reverse Perfect draw occurs. Find the probability of this event and compute the HAs corresponding to several plausible sets of payoff odds. What payoff gives the smallest positive HA?

16. An alternate version of 8-ball Rouleno eliminates excluded eight-number permutations and designates the 8 ball as the house ball. Single-number bets on the number 8 are not permitted, and if the 8 ball is drawn last, all bets lose unless the drawing is Perfect, in which case all single-number bets and Perfect bets win. Find the HA of a single-number bet on a number from 1–7 paying 7–1.

Chapter 4

Card Games

The standard 52-card deck contains 13 cards: ace (1) through 10, jack, queen, and king, in each of 4 suits: hearts ($\heartsuit$), clubs ($\clubsuit$), diamonds ($\diamondsuit$), and spades ($\spadesuit$). Clubs and spades are black; hearts and diamonds are red. (Televised and online poker displays sometimes show cards onscreen using green clubs and blue diamonds for enhanced clarity in a small viewing area; hearts and spades retain their original colors.) The ace can be regarded as either low or high, depending on the rules of the game in question. Some decks may include one or more jokers, which have no suit or rank, and often function as wild cards. Casino games use anywhere from 1–8 decks shuffled together; when four or more decks are used, they are dealt from a box called a *shoe*. (Section 4.7 considers game variations that use nonstandard decks of various types: with different numbers of suits or ranks within a suit.)

When considering probabilities involving card games including blackjack, baccarat, and several of the games examined in this chapter, unless the cards are completely reshuffled between hands, a card dealt in one hand cannot appear again until the cards are shuffled and returned to play. This means that successive hands depend on the cards previously dealt, which makes the mathematics more interesting and also may provide opportunities for players to gain an edge over the casino. Successive hands of blackjack or baccarat are dependent; this is in contrast with successive rolls of dice or spins of a wheel, which are independent. Each die roll or wheel spin is a fresh start: the outcome is not affected by any previous rolls or spins.

A simpler analysis, which may be accurate enough for most purposes, can sometimes be performed by assuming that successive hands are independent, where each card dealt has no effect on subsequent probabilities. This is called the *infinite deck approximation*, as it is effectively assuming that the game is dealt from a shoe containing infinitely many decks, so that the probability of dealing a certain card is constant throughout the game.

4.1 Poker

Poker is the name given to a wide variety of card games in which players compete to hold the highest-ranking hand when dealt 3–7 cards. Two particularly popular forms of poker in casinos are *5-Card Draw*, which is the basis for most versions of video poker, and *Texas Hold 'Em*, which has grown in prominence with the rise of televised poker in the 21st century.

In a hand of 5-card draw, players are each dealt 5 cards, and a round of betting follows. Following that round, players who have not folded have the option to discard some of their cards and receive replacements, after which a second round of betting takes place. Once all players have either folded or contributed an equal amount to the pot, the hands are revealed, and the player with the best hand wins the pot.

Texas Hold 'Em features more betting and the added feature that players make their best 5-card hand from 7 cards. Two of these are a player's individual hole cards, dealt to each player at the start of a hand and kept concealed. The other 5 are *community cards*, sometimes called the *board*, which are dealt face up to the center of the table and may be used by all players to build a 5-card hand. There are four rounds of betting in Texas Hold 'Em:

- One round after the players are dealt their hole cards.

- Three community cards, the *flop*, are then dealt, and a second round of betting follows.

- The fourth community card, the *turn*, is dealt, and the third betting round follows it.

- The final community card, the *river*, is dealt, and the last round of betting takes place.

As in 5-card draw poker, the winner at the end of the showdown, when the players who have not yet folded reveal their cards, is the player still in the game who has the highest-ranked hand. In practice, skilled players will fold weak hands before betting very much money, so a showdown might involve only 2 or 3 players from a full table of 8 or 9.

Both games use the same hand ranking. Table 4.1 shows the ranking of 5-card poker hands from highest to lowest.

Ties among hands of the same rank are broken, if possible, by looking at the highest-ranked card, so a king-high straight will beat a queen-high straight, and 3 aces outranks 3 jacks. Suits have no standing in most forms of poker; the straight flush ♡98765 and the straight flush ◇98765 tie in a showdown.

Owing, of course, to the structure of betting in poker, the winner in a given hand is not necessarily the best hand as ranked by Table 4.1, but rather the best hand *that is still in the game at the showdown*. A hand folded prior to the end of betting is a losing hand, even if it would have been the winner had its owner not folded. This suggests the classification of many forms of poker as skill games rather than games of pure chance.

TABLE 4.1: Five-card poker hands

Hand	Description
Royal flush	AKQJ10 of the same suit
Straight flush	5 cards of the same suit and in sequence
Four of a kind	Four cards of the same rank
Full house	3 of a kind and a pair
Flush	5 cards of the same suit, not in sequence
Straight	5 cards in sequence, not all the same suit
Three of a kind	3 cards of the same rank
Two pairs	Two pairs of cards of different ranks
One pair	Two cards of the same rank
High card	Five cards of different ranks, not all the same suit

Variations with Multiple Wagers

Some carnival games based on poker attempt to mimic the flow of a standard poker game by incorporating multiple rounds of betting into game play. We consider those in this section. These games typically require an *ante* bet from each player before any cards are dealt; subsequent wagers are frequently required to be a fixed multiple of the ante.

Caribbean Stud Poker

Deck composition:	52 cards
Hand size:	5
Community cards:	None
Betting rounds:	2

One of the most successful carnival games based on poker was *Caribbean Stud Poker* (CSP), a variation on 5-card stud poker. Stud poker differs from draw poker in that players have no opportunity to exchange dealt cards for replacements; each player must play their original hand. While CSP's origins are unclear, it is certain that the game was sold to game developer Mikohn for $30 million, setting a standard for success to which carnival game developers have since aspired [45].

A round of CSP sees 5 cards dealt to each player and to the dealer. The game is played by each gambler individually against the dealer. Unlike in Texas Hold 'Em, players do not compete against each other. One of the dealer's cards is dealt face up to aid players in their wagering decision. Players make an initial Ante bet before cards are dealt. After seeing their hands and the dealer's upcard, they may choose to fold, forfeiting their ante, or back up the ante with a Call bet of double the ante in the hope that their hand will outrank the dealer's.

With no further rules, CSP would be an even game, with no house advantage. The casino derives an edge from how the hands are compared. The dealer's hand is said to *qualify* if it contains at least an ace and a king, including any pair or higher. If the dealer fails to qualify, the player's Ante bet is paid even money while the Call bet pushes. If the dealer qualifies, player hands that beat the dealer's receive even money on the Ante, and the Call bet is paid in accordance with Table 4.2.

TABLE 4.2: Caribbean stud poker payoffs [6]

Player Hand	Payoff odds
Royal flush	100 to 1
Straight flush	50 to 1
Four of a kind	20 to 1
Full house	7 to 1
Flush	5 to 1
Straight	4 to 1
Three of a kind	3 to 1
Two pairs	2 to 1
One pair	1 to 1
AK high	1 to 1

There are 1,135,260 possible dealer hands that do not qualify [37], so the probability of a dealer qualifying hand is

$$\frac{1,463,700}{2,598,960} \approx .5632.$$

By requiring that the dealer have a qualifying hand, some potential high-paying player hands win only the ante bet if the dealer doesn't qualify. A player making a $10 ante bet and backing it up with a $20 call bet upon being dealt a straight, for example, wins only $10 instead of $90 if the dealer's hand doesn't qualify. If you beat the low probability of .00392 by receiving a dealt straight, you want to cash in for as much as you can, but the requirement that the dealer qualify takes some of that opportunity away.

This rule shows the true challenge facing a CSP player. You lose if the dealer's hand is good enough to beat yours, as in traditional stud poker, but you also lose, in the sense of winning less than you ought to, if the dealer has a bad hand.

Example 4.1. If you are dealt

$$6\clubsuit \ Q\clubsuit \ K\spadesuit \ 9\clubsuit \ 6\diamondsuit,$$

then you should make the Call bet regardless of the dealer's upcard, since your pair of 6s is a winning hand according to Table 4.2. If the dealer qualifies with a hand you can beat, you win $3; if the dealer qualifies but beats your pair, you lose $3. If the dealer fails to qualify, with probability approximately .4368, you will win $1. ∎

Example 4.2. However, the hand

$$A\diamondsuit\ K\diamondsuit\ 4\heartsuit\ 3\heartsuit\ 2\heartsuit,$$

which is the lowest possible AK-high hand, might not justify a Play bet. If the dealer is showing the $K\spadesuit$, his or her chance of qualifying is increased past 56.32%. The expectation if you fold your hand is −$1; if you make the Call bet, we know that

$$E < (1) \cdot .4638 + (-3) \cdot .5632 = -\$1.2258,$$

so folding is the more prudent option.

This decision really made no use of the dealer's upcard, since the value −$1.2258 is an upper bound for your expectation. Folding $A\diamondsuit\ K\diamondsuit\ 4\heartsuit\ 3\heartsuit\ 2\heartsuit$ against any dealer upcard is the better choice. ∎

Careful analysis shows that an effective player strategy for CSP totally disregards the dealer's card and simply calls for the player to call with any hand of AKJ83 or higher. This gives the casino an advantage of 5.22% [35].

Caribbean Stud Poker was at one time one of the most popular carnival games in Las Vegas, but its popularity has diminished over time due to competition from other poker-based carnival games.

Four-Card Poker

Deck composition:	52 cards
Hand size:	5 cards dealt, 4 used
Community cards:	None
Betting rounds:	2

A number of carnival games based on poker are played with fewer than 5 cards in each hand. This changes hand frequencies and the relative rank of standard poker hands. *Four-card poker* (4CP) pits the gambler against the dealer in a head-to-head game matching four-card poker hands. In addition to the usual practice of paying off winning hands at less than true odds, as was seen in CSP, the casino derives an additional advantage from the rule that the player gets five cards to construct a four-card poker hand, while the dealer gets six. One of the dealer's cards is dealt face up as an aid to players in judging the hand's strength.

Betting in 4CP occurs in several stages, just as in CSP:

• The player makes an initial Ante bet before the cards are dealt.

- Upon seeing his or her cards and forming the best possible four-card hand, the player may fold, forfeiting the Ante bet, or may make an additional Play bet ranging from one to three times the Ante bet.

- An optional Aces Up bonus bet, made before the deal, pays off in accordance with Table 4.3 if the player's hand is a pair of aces or higher, regardless of the dealer's hand.

TABLE 4.3: Four-Card Poker: Aces Up bonus bet pay table

Hand	Payoff odds
Four of a kind	50–1
Straight flush	40–1
Three of a kind	8–1
Flush	5–1
Straight	4–1
Two pairs	3–1
Pair of aces	1–1

The player and dealer hands are compared, and the higher hand wins with the player taking ties. The payoff is even money on both the Ante and Play bets. At the Soaring Eagle Casino, an automatic bonus is paid on the Ante bet if the player's hand is three of a kind or higher, even if the hand is beaten by the dealer. Table 4.4 contains the payoffs.

TABLE 4.4: Four-Card Poker: Automatic Bonus pay table

Four of a kind	25–1
Straight flush	20–1
Three of a kind	2–1

Example 4.3. Tables 4.3 and 4.4 indicate that four of a kind is better than a straight flush in 4CP, which is the opposite order from five-card poker. Careful counting will confirm this.

There are 13 possible four of a kinds, one for each rank. Any one of them may be dealt with any of the remaining 48 cards as the fifth card. This gives a total of $13 \cdot 48 = 624$ ways to draw four of a kind. This is exactly the number of four-of-a-kind hands in standard five-card poker.

Four-card straight flushes are more common than their five-card counterparts. As with five-card poker, we consider the cards comprising the flush in order. Note that the royal flush is not broken out as a separate hand, so any card from ace through jack may be the lowest card in a four-card straight flush, for a total of 44 distinct straight flushes. As with four of a kinds, the fifth card in the player's hand may be any of the remaining 48 *except* the one card that completes a higher straight flush.

For example, consider the straight flush 5678♣. If the fifth card dealt to the player is the 9♣, then the four-card hand will be played as 6789♣. All straight flushes except those starting with a jack (a total of 40) require that this deletion occur. We have

- 40 four-card straight flushes whose lowest card is ace through 10. Each can be combined with 47 other cards in the deck.

- 4 four-card straight flushes consisting of a suited JQKA. These can be filled out to a 5-card hand by any of the other 48 cards.

This gives
$$40 \cdot 47 + 4 \cdot 48 = 44 \cdot 48 - 40 = 2072$$
four-card straight flushes, over three times as many straight flushes as four of a kinds. ∎

These results are for the player's hand; Exercise 1 asks for the corresponding calculations—and the same conclusion—for the dealer's hand.

As with CSP, 4CP comes with a major strategy decision: whether or not to make the Play bet based on one's cards. Unlike CSP, there is no question of the dealer qualifying or not, so the only question to be answered is whether or not your hand has a good chance of beating the dealer's.

The Play bet pays off at even money, so you should make that bet if your hand falls into the top half of all four-card poker hands. The rack card (a card available at the table that explains the rules, and possibly strategies, for a game) at the Soaring Eagle Casino states that you should make the Play bet if your hand is a pair of 3s or higher. Is this sound advice? What is the "beacon hand" that separates hands worth a Play bet from those which should be folded?

Since the player is dealt five cards to make his or her best four-card hand, we shall consider five-card combinations, of which there are 2,598,960. How many of these hands lack even a pair when reduced to their best four cards? As with five-card hands, the standard procedure is to compute the number of higher hands and subtract that total from this number. The hand distribution for 4CP hands is given in Table 4.5.

Our beacon hand would be hand #1,299,480, which is still among the list of high-card hands. While the Soaring Eagle's advice is valid, there are a number of hands—pairs of 2s as well as some high-card hands—that should be played rather than folded. Of course, if your high-card hand cannot beat the dealer's upcard—as when your hand is king-high and the dealer shows an ace, for example—you should fold immediately. This is one situation where the value of the dealer's upcard is useful as a guide to gameplay.

The apparently important question of how big a Play bet to make is, in practice, quite simple to answer. When you make the Play bet, you are doing

TABLE 4.5: Four-Card Poker: Hand distribution

Hand	Frequency
Four of a kind	624
Straight flush	2072
Three of a kind	58,656
Flush	114,616
Straight	101,808
Two pairs	123,552
One pair	1,098,240
High card	1,302,540

so because you expect that your hand will win. If this bet is worth making, it is worth making for the maximum possible value: three times your Ante bet. Any smaller Play bet is leaving expected winnings on the table.

Three Card Poker

Deck composition:	52 cards
Hand size:	3 cards
Community cards:	None
Betting rounds:	2

If five-card poker hands are too complicated, perhaps restricting your hand to three cards might be more to your liking. Apparently enough people feel this way to make *Three Card Poker* (3CP) another carnival game that's found a place in the game lineup at a number of casinos.

The title gives the key to the game: players receive three cards apiece, and match their cards up against the dealer's three-card hand—once again, this is a game where players face off against the dealer instead of each other. 3CP is very much like Caribbean Stud Poker in that players make an Ante bet and then can choose, based on their cards, whether or not to fold or make a second "Play" bet. Also coming over from CSP is the notion of a qualifying hand for the dealer, which here is any hand holding at least a queen. One difference between CSP and 3CP is that none of the dealer's cards is exposed to help players make the decision whether or not to fold. If the dealer fails to qualify, Ante bets pay even money and Play bets push. If the dealer qualifies, winning player hands pay even money on both the Ante and Play bets; if the dealer's hand is higher, then both bets lose. Additionally, the Ante bet pays a bonus if the player's hand is a straight or higher, even if it is beaten by the dealer.

Example 4.4. What is the probability of the dealer qualifying?

There are $\binom{52}{3} = 22,100$ three-card hands, and $\binom{40}{3} = 9880$ of them contain no queens, kings, or aces. From this latter number, of course, we must remove any hand containing a pair or higher.

For example: Any card from 4 to J can be the highest card in a three-card straight flush; these hands run from 432 through J109. There are 8 ranks and 4 suits to consider, for a total of 32 such hands. These "exceptional" hands are counted in Table 4.6.

TABLE 4.6: Three Card Poker: Qualifying hands with no AKQ

Hand	Count
Straight flush, J109 or lower	32
Three of a kind, jacks or lower	40
Flush, jack-high or lower	$4 \cdot \binom{10}{3} - 32 = 448$
Straight, J109 or lower	$32 \cdot 4 \cdot 4 - 32 = 480$
Pair, jacks or lower	$10 \cdot \binom{4}{2} \cdot 36 = 2160$

Adding up in this table gives 3160 possible qualifying hands with no card higher than a jack, and subtracting gives 6720 nonqualifying hands. The probability that the dealer qualifies is therefore

$$1 - \frac{6720}{22,100} = \frac{15,380}{22,100} \approx .6959,$$

so the dealer qualifies about 70% of the time. ∎

Armed with this information, we look next at the pay table for player hands that beat a qualifying dealer hand. The Ante Bonus pay table varies among casinos [95]; a common payoff structure is 1 to 1 on straights, 4 to 1 on three of a kinds, and 5 to 1 on straight flushes. Some casinos recognize a "Mini Royal Flush": a suited AKQ, and reward it with a 50 to 1 Ante Bonus payoff.

There is no pay table for the Play bet—all bets are paid at 1 to 1—but there is an optional Pair Plus bet that pays off player hands of at least a pair at odds. This bet must be made with the Ante bet before the cards are dealt, of course. Pay tables can vary among casinos; one of the most common is Table 4.7. It should be noted that the Pair Plus bet is paid off even if the player's hand does not beat the dealer's or if the dealer fails to qualify.

TABLE 4.7: Three Card Poker payoffs: Pair Plus bet [95]

Hand	Payoff
Straight flush	40 to 1
Three of a kind	30 to 1
Straight	5 to 1
Flush	4 to 1
Pair	1 to 1

As we might expect, these payoffs are far below the probabilities of the various hands. In the case of a straight flush, there are 48 possible hands (any card except a deuce can be the high card in a three-card straight flush), and so

$$P(\text{Straight flush}) = \frac{48}{22,100} \approx \frac{1}{460.42}.$$

When considering CSP, we identified the lowest hand justifying a call bet, AKJ83. A similar analysis can be applied to 3CP; once again we seek the minimum hand for which the expectation is greater than the –$1 outcome achieved by folding and forfeiting the Ante bet. The "beacon hand" for 3CP turns out to be Q64: any player hand better than this should be backed with a Play bet and any lesser hand should be folded. Following this one piece of advice limits the casino's advantage to a mere 3.37%—actually quite reasonable for a table game. If instead you choose to "mimic the dealer" by calling on every hand holding at least a queen, that gives the casino a 3.45% edge—not much of a change. The nonstrategy of always calling, on the other hand, raises the HA to 7.65%: more than double that when using the beacon hand strategy above [95].

Two-Card Joker Poker

Deck composition:	54 cards: 2 jokers
Hand size:	2 cards
Community cards:	None
Betting rounds:	2

Two-Card Joker Poker (2CJP) is a stripped-down version of poker that deals 2-card hands, with 2 jokers added to the deck as semi-wild cards that can be used to form a pair. With only two cards per hand, the rank of poker hands is different than in 5-card poker. Table 4.8 shows the hands in order from highest to lowest. Aces rank high, but an ace may be used with a deuce to form a straight or straight flush [53].

TABLE 4.8: Two-Card Joker Poker hands

Hand	Description
Two jokers	Both jokers
Royal flush	Suited ace and king
Straight flush	Two suited cards in sequence, other than AK
Straight	Two cards of different suits in sequence
Pair	Two cards of the same rank, possibly with one joker
Flush	Two cards of the same suit, not in sequence
High card	Two unmatched cards of different suits

The rankings state that a straight outranks a flush in 2CJP; this is the reverse of their order in 5-card poker. To confirm this order, we note up front that there are 52 straight or royal flushes, since any card can be the lower card. These must be removed from the count of both straights and flushes. Jokers need not be considered here since they may not be used to complete a straight or flush.

The number of flushes is $4 \cdot \binom{13}{2} - 52 = 260$. Straights may be counted by picking the lower card, which may be done in 52 ways, and then selecting any of the three cards that are next in rank without completing a straight flush. This gives $52 \cdot 3 = 156$ straights. Straights are only 60% as likely as flushes and so rank higher.

Example 4.5. A count of pairs must account for the jokers. Define a *jokered* pair as a pair containing a joker and a *natural* pair as a pair without a joker. The probability of drawing a pair in 2CJP is

$$\frac{\text{\# of jokered pairs} + \text{\# of natural pairs}}{\binom{54}{2}} = \frac{52 \cdot 2 + 13 \cdot \binom{4}{2}}{1431} = \frac{182}{1431} \approx .1272,$$

slightly better than 1 chance in 8. ∎

The numerator in this calculation confirms that pairs are less common than flushes in 2CJP, as Table 4.8 shows.

Betting occurs in two rounds: an Ante bet placed before any cards are dealt, and an optional Call wager made after the player has seen his or her cards and decides whether or not to continue. A player dealt a weak hand may forgo the Call bet and fold, forfeiting the Ante bet. As in 3CP, the dealer must qualify with a hand holding at least a queen for Call bets to have any action; if the dealer fails to qualify, Ante bets pay 1–1 while Call bets push.

If the player's hand beats a qualifying dealer hand, the Ante bet pays 1–1 and the Call bet is paid in accordance with Table 4.9.

TABLE 4.9: Two-Card Joker Poker: Call bet pay table [53]

Player's winning hand	Payoff
Two jokers	8–1
Royal flush	5–1
Straight flush	3–1
Straight or lower	1–1

Adding up the values already computed or easily determined for hands in 2CJP shows that 780 of the possible 1431 hands are high-card hands, with no pairs, straights, or flushes. This information allows an easy calculation of the dealer's chance of qualifying. Of the 780 high-card hands, 120 are ace-high:

• There are 4 choices for the ace.

- There are then 30 choices for the other card: it cannot be another ace (3 cards), a joker (2), a king or deuce that would complete a straight (8), or a 3 through queen of the same suit as the ace that would complete a flush (10). This removes 23 cards, leaving 30 choices for the second card.

Similar analysis shows that there are 120 king-high hands and 108 queen-high hands, for a total of 348 high-card hands with a queen or better. The probability that the dealer qualifies with a queen-high or better hand is then

$$\frac{1 + 4 + 48 + 156 + 182 + 260 + 348}{1431} = \frac{999}{1431} \approx .6981,$$

so the dealer qualifies just under 70% of the time, as was the case in 3CP.

Casino War

Deck composition:	52 cards
Hand size:	1 card
Community cards:	None
Betting rounds:	1 or 2

One-card poker, anyone?

On the face of it, that seems like an unlikely game—after all, with only one card, every hand is a straight flush.

In the 1997 movie *Vegas Vacation*, one scene is set in a seedy-looking casino portrayed by the Klondike Casino on the Las Vegas Strip, which closed in 2006. This casino offers a number of nonstandard games of chance, including a version of the children's card game War. *Casino War*, a gambler's version of War which debuted in 1994 in northern Nevada, might reasonably be considered as a one-card poker game [104]. Player and dealer are each dealt one card from a six-deck shoe (the number of decks may vary), and the high card wins. Aces are always high. At this point, the game is even—both sides have an equal chance of drawing the higher card.

The casino's edge comes from how ties are handled. If the two cards match, players may either *surrender*, forfeiting half their wager and ending the game right away, or, as in the children's game, *go to war*. If war is declared, the player must double his or her bet, and the dealer then burns three cards (removes them from play without exposing them) and deals a second card to each side. If the player's card is higher, the player wins the second bet, which is paid off at 1 to 1 odds, and the first bet is a push; if the dealer's card is higher, the player loses both bets.

A tie on the second card is handled differently at different casinos. A common resolution is to pay the player a bonus equal to the amount of the original bet while the actual bets are declared pushes, a net win of 1 unit. On a $1 bet, then, we have the following player outcomes:

Result	Win/Loss
Win without war	+$1
Lose without war	−$1
Surrender	−$.50
Win after war	+$1
Lose after war	−$2
Tie after war	+$1

The source of the casino's advantage is clear from this chart: In the event of a casino win after a tie, 2 units are won, while the player can never win more than the amount of his or her original wager even if 2 units are at risk in a war.

Example 4.6. Casino War also offers a Tie bet, which pays 10 to 1 if the first two cards match. Assuming a single hand dealt from a fresh shoe, the HA for this bet depends on the number of decks in play. We shall compute this house edge for a variable number of decks, denoted by n; substitution of commonly used values of n will then give the HA appropriate to any particular game.

Once the first card is drawn, we seek the number of cards remaining in the shoe that match its rank. In an n-deck game, there are $4n - 1$ such cards. Since the entire shoe holds $52n$ cards and one (the player's card) has been removed, the probability of a match is

$$p = \frac{4n - 1}{52n - 1}.$$

For a single-deck game, $p = \frac{1}{17}$; as n increases, this probability approaches $\frac{4}{52}$, or $\frac{1}{13}$. The casino's advantage comes from paying 10 to 1 on what is no better than a 12 to 1 shot. The exact HA may be found from the equation

$$E(n) = (10) \cdot \left(\frac{4n - 1}{52n - 1} \right) + (-1) \cdot \left(1 - \frac{4n - 1}{52n - 1} \right) = -\frac{8n + 10}{52n - 1}.$$

The expected values are tabulated for the various values of n typically used in casinos in Table 4.10.

TABLE 4.10: House edge on the Tie bet for Casino War

# of decks	HA
1	35.29%
2	25.24%
4	20.29%
6	18.65%
8	17.83%

It is clear that you should not make this bet, but it is of interest to notice that the casino's edge actually decreases with the number of decks in the shoe, which is the opposite of the effect in a game like blackjack. Increasing the number of decks past eight (the maximum typically in use in casinos) cannot, however, eliminate the casino advantage completely. In the limit as $n \to \infty$, the expectation approaches

$$E = (10) \cdot \left(\frac{1}{13}\right) + (-1) \cdot \left(\frac{12}{13}\right) = -\frac{2}{13},$$

which corresponds to approximately a 15.38% house edge. ∎

Example 4.7. It may be stretching the definition of the word to speak of a "strategy" for a game as simple as Casino War, but the question of whether to surrender or go to war after a tie is a place where player choice is involved, and thus a place where an optimal strategy may be determined.

If you surrender a \$1 bet, your expectation is −\$.50. If you go forward with the war, then the probability of winning is again dependent on the number of decks in use. You will either win \$1, whether through winning the war or tying, or lose \$2 when the war is lost. Let p be the probability of a tie; it follows that

$$P(\text{Win } \$1) = \frac{1}{2} + p$$

and

$$P(\text{Lose } \$2) = \frac{1}{2} - p.$$

We assume no knowledge about the cards remaining to be dealt, and again assume that we're starting at the top of a fresh n-deck shoe. Three cards are known at the start of the war: the two matching cards that triggered the war and the player's second card. The burn cards are not exposed when dealt, so we ignore them—the result will be another long-term average value that is suitable for quick calculations like this one. Two cases emerge:

- If your second card matches the first two in rank, then the conditional probability of a fourth card of that rank falling to the dealer is

$$p_1 = \frac{4n - 3}{52n - 3}.$$

  This case has probability

$$q_1 = \frac{4n - 2}{52n - 2}.$$

- If your second card is of a different rank than the first two, then the conditional probability of a second match is

$$p_2 = \frac{4n - 1}{52n - 3},$$

  and this case has probability

$$q_2 = \frac{48n}{52n - 2}.$$

The probability p of a tie and the expectation E are both functions of n. We find that

$$p(n) = q_1 p_1 + q_2 p_2 = \frac{208n^2 - 68n + 6}{(52n - 2)(52n - 3)},$$

where n is the number of decks in the shoe. Hence, your expectation as a function of n is

$$E(n) = (1) \cdot \left(\frac{1}{2} + p(n)\right) + (-2) \cdot \left(\frac{1}{2} - p(n)\right) = \frac{-36 \cdot (26n - 3)}{(26n - 1)(52n - 3)} - \frac{7}{26}.$$

For the commonly used values of n, Table 4.11 contains the expectation for the player who chooses to go to war.

TABLE 4.11: Expected return when going to war, Casino War

n	$E(n)$
1	-\$.321
2	-\$.296
4	-\$.282
6	-\$.278
8	-\$.276

Once again, a limiting value for the case of infinitely many decks can be computed, and here that limit is $-\$\frac{7}{26} \approx -\$.269$.

Since the expected value for any number of decks is greater than −50¢, it follows that you should *never* surrender, and thus that the casino derives an additional advantage whenever a player surrenders. ∎

This is not a principle limited to Casino War: In any casino game where player choice is possible, a player who makes a non-optimal choice works against his or her own interests and increases the casino's advantage.

Dueling for Dollars

Deck composition:	52 cards
Hand size:	1 card
Community cards:	1
Betting rounds:	1 or 2

Dueling for Dollars is a product of Galaxy Gaming. This variation on Casino War allows gamblers to bet on either their card or the community card dealt to the dealer. The basic bets in Dueling for Dollars are the same as for Casino War. Ties on the second card, after going to war, are resolved in favor of the gamblers: if the War cards tie, all players win on their Play bet and push on their Ante bet, whether they've bet on their card or the dealer's card [113].

Dueling for Dollars extends Casino War with optional side bets. As in Casino War, the *Tie* bet pays off if the player's card and dealer's card have the same rank, but the payoffs are better than 10–1. Several pay tables are available as a casino manager might select; they are shown in Table 4.12.

TABLE 4.12: Payoff odds: Dueling for Dollars Tie bet [113]

Pay Table	Suited pair	Unsuited Pair
A	12–1	12–1
B	18–1	10–1
C	20–1	10–1

In determining which pay table gives the players the best game, we can instantly discount table B, which is always beaten or tied by table C. As in Casino War, expected values of this Tie bet depend on the number of decks in play. Suppose that a given pay table pays X to 1 on a suited pair and Y to 1 for an unsuited pair. (Pay table A covers the case where $X = Y$.) If n is the number of decks used, the probability of a suited pair is

$$\frac{n-1}{52n-1}$$

and the probability of an unsuited pair is

$$\frac{3n}{52n-1},$$

assuming that the hand is dealt from a full shoe. The expected value of a $1 tie bet is

$$E = (X) \cdot \frac{n-1}{52n-1} + (Y) \cdot \frac{3n}{52n-1} + (-1) \cdot \frac{48n}{52n-1} = \frac{n \cdot (X + 3Y - 48) - X}{52n-1}.$$

Common values for n are 5, 6, and 8; these give the HAs shown in Table 4.13. Looking at the HAs for pay tables A and C reveals that pay table C gives the

TABLE 4.13: Dueling for Dollars Tie bet: Comparing pay tables

Pay Table	5 decks		6 decks		8 decks	
	E	HA	E	HA	E	HA
A	–$0.0463	4.63%	–$0.0386	3.86%	–$0.0289	2.89%
B	–$0.0695	6.95%	–$0.0579	5.79%	–$0.0434	4.34%
C	–$0.0386	3.86%	–$0.0257	2.57%	–$0.0096	0.96%

player the best game.

Another Dueling for Dollars option is the *Two-Card Poker Bonus* side bet, which is a bet on the 2-card poker hand formed by the dealer and player cards; this pays off if the 2 cards form a flush, pair, straight, or straight flush. Since Dueling for Dollars is dealt from multiple decks and uses no jokers, the probabilities of these various hands differ from those used in Two-Card Joker Poker. As with the Tie bet, we let n be the number of decks. Once the first card is dealt from a fresh shoe, we have the following probabilities:

$$P(\text{Straight flush}) = \frac{2n}{52n - 1}$$
$$P(\text{Straight}) = \frac{6n}{52n - 1}$$
$$P(\text{Flush}) = \frac{10n}{52n - 1}$$
$$P(\text{Pair}) = \frac{4n - 1}{52n - 1}.$$

One possible pay table is shown in Table 4.14.

TABLE 4.14: Dueling for Dollars Two-Card Poker Bonus pay table [113]

Hand	Payoff odds
Straight flush	2–1
Straight	1–1
Flush	1–1
Pair	2–1

The HA on this bet with Table 4.14 runs in a very narrow range: from 4.63% in a 5-deck game down to 4.34% if 8 decks are used.

Mississippi Stud

Deck composition:	52 cards
Hand size:	5 cards
Community cards:	3
Betting rounds:	4

Mississippi Stud is a 5-card stud poker variation in which players can risk up to 10 times their original Ante bet. Gamblers play against a pay table, not each other or the dealer. After making the initial Ante bet, every player receives 2 cards and 3 community cards are dealt face down to the table. Players may either fold or make the 3rd Street bet of 1–3 times their Ante, after which the first community card is revealed. Players may now fold or make the 4th Street bet, which again can be 1–3 times the Ante. The second community card is turned up, and players either make a 5th Street bet of 1–3 times their Ante, or fold. At this point, players who remain active have 4–10 times their original wager in play.

The final community card is revealed, and any player whose hand is at least a pair of jacks or better wins. Players holding a pair of 6s through 10s push all bets. All bets are paid at fixed odds. Pay tables for Mississippi Stud vary slightly among casinos; one pay table is shown in Table 4.15.

TABLE 4.15: Mississippi Stud pay table

Hand	Payoff odds
Royal flush	500–1
Straight flush	100–1
4 of a kind	40–1
Full house	10–1
Flush	6–1
Straight	4–1
3 of a kind	3–1
Two pairs	2–1
Pair, jacks or better	1–1
Pair, 6s through 10s	Push

Where there are choices in a casino game, it should be possible to develop an optimal strategy, and mathematician Joseph Kisenwether has devised such a strategy for Mississippi Stud [78]. The strategy involves assigning point values to cards:

- High cards: Face cards and aces count 2 points each.
- Middle cards: 6s through 10s count 1 point.
- Low cards: 2s through 5s count 0 points.

An original hand of $Q\heartsuit$ $4\diamondsuit$ counts 2 points, for example. These point values are connected to the payoff value of a pair of cards of that rank.

The best action for each street bet is based on the value of the known cards, with some exceptions made for hands with high potential.

1. On the 3rd Street bet:

   - **Bet 3× the ante on a pair.** A high or middle pair is guaranteed to be a nonlosing hand, so there's no risk in backing it up with the maximum possible wager. Low pairs also have potential for improvement and are worth maximum support.

   - **Bet 1× the ante on a hand totaling 2 points or more, or on a suited 5 and 6.** These hands have some potential to improve, but you should hold back on a 3× bet until you've seen them improve on the first community card.

   - **Fold all other hands.** The probability of a hand like $2\spadesuit$ $8\heartsuit$ improving to at least a middle pair is too small to risk more money.

One important strategic principle for poker players, especially stud poker players, is due to gaming expert John Scarne [89, p. 587]:

When you have nothing, get out.

This advice strongly encourages folding weak hands rather than putting more money at risk on a hand with minimal chance of improvement. While this is perhaps most appropriate for standard stud poker games when players can see other players' cards and compete directly against each other, it is a sound idea in any poker game.

2. On 4th Street:

- **Bet 3× the ante on any pair of 6s or higher (a guaranteed non-losing hand), 3 cards of a royal flush, or 3 cards to a straight flush meeting any of the following criteria.**

   - No gaps with the lowest card a 5 or higher, such as $5\diamondsuit$ $6\diamondsuit$ $7\diamondsuit$.
   - One gap and a high card. For example, $J\heartsuit$ $9\heartsuit$ $8\heartsuit$.
   - Two gaps and two high cards. For example, $K\clubsuit$ $J\clubsuit$ $9\clubsuit$.

   All of these 3-card hands have the potential to be completed to a straight or flush as well as a straight flush. Separating out these hands isolates the straight flush draws which have the best potential to complete to a hand with a high or middle pair.

- **Bet 1× the ante on any of the following:**

   - Any 3 cards of the same suit.
     Among other hands, this covers the straight flush draws where a 3× bet is inadvisable, such as $2\heartsuit$ $3\heartsuit$ $5\heartsuit$.
   - A pair of 2s through 5s.
     These low pairs are not winning hands yet, but might catch a 3rd card of the pair's rank and so advance to a win.
   - Any 3-point hand, for example, $A\heartsuit$ $9\heartsuit$ $3\spadesuit$.
     Hands like this, which have at least 2 high or middle cards, have potential to catch a paying or break-even pair on the final two community cards.
   - A straight draw with no gaps if the lowest card is a 4 or higher, for example, $4\diamondsuit$ $5\spadesuit$ $6\spadesuit$.
     The 3-card straights 234 and 345 are weaker hands because they contain no middle or high cards. Additionally, a 234 hand is limited at the low end. While 345 can complete to the three straights A2345, 23456, and 34567, a 234 only admits A2345 and 23456 as successful completed straights.
   - A straight draw with one gap and 2 middle cards or better, for example, $5\clubsuit$ $7\heartsuit$ $8\spadesuit$.

Again, these last two categories separate out hands with some potential to form a straight and a backup possibility of a middle or high pair.

- **Fold everything else.**

3. On 5th Street:

- **Bet 3× the ante on any pat hand, a flush draw with 4 cards of the same suit, or any outside straight draw.**

  An outside straight draw consists of four cards in sequence that can be completed to a straight on either end. A 4567 hand is an outside straight draw; a 4678 hand is not, since only a 5 will fill out the straight.

  Raising pat hands as much as possible is an obvious strategy. There are 9 cards that will fill out a 4-card flush to a 5-card flush, and 8 cards that will complete an outside straight. Measured against the potential for a high payoff, a 3× raise on these hands is worth the risk.

- **Bet 1× the ante on an inside straight draw such as the 4678 mentioned above, a low pair, a hand worth 4 points, or a hand with 3 mid-ranked cards that was raised on an earlier street.**

  These hands have enough potential for improvement to make it worth staying in the game, but their high probability of winding up as losing hands prompts the advice not to go all the way to a 3× raise.

- **Fold all other hands.**

Note that, although the game rules allow 2× ante bets, those should never be made when following the optimal strategy: either bet 1×, bet 3×, or fold. Playing Mississippi Stud using optimal strategy gives the casino a 4.91% advantage [78].

Example 4.8. Suppose that your original two cards were $Q\heartsuit\ 4\diamondsuit$. As a 2-point hand, this justifies a 1× raise on 3rd Street. If the first community card is the $K\diamondsuit$, the hand is now worth 4 points, so you should make a 1× bet on 4th Street. When the second community card is the $A\clubsuit$, the hand is worth 6 points, thus justifying a further 1× bet on 5th Street.

These 1× raises are made in the hope that the third community card will complete a pair of queens, kings, or aces, and so qualify the hand for a 1–1 payoff on the $4 wagered. Nine of the remaining cards will pair one of the high cards, so probability of this win is

$$\frac{9}{48} = .1875.$$

Folding this hand on 5th Street results in a loss of $3. Making the final $1\times$ bet gives the hand an expected value of

$$E = (4) \cdot \frac{9}{48} + (-4) \cdot \frac{39}{48} = -\$\frac{120}{48} = -\$2.50,$$

which is a smaller loss than $3, justifying the 5th Street bet. ∎

Mississippi Stud also offers an optional 3-Card Bonus bet on the composition of the 3 community cards, which is made before any cards are dealt. If a player chooses to make this bet, it remains in action even if the player folds his or her hand. If the community cards contain at least a pair, the bet pays off like the main bet: according to a fixed pay table. The pay table from the Greektown Casino in Detroit, Michigan is shown in Table 4.16.

TABLE 4.16: Mississippi Stud: 3-Card Bonus pay table

Hand	Payoff odds
Straight flush	40–1
3 of a kind	30–1
Straight	6–1
Flush	3–1
Pair	1–1

Unlike the main Mississippi Stud bets, there is no need for strategic choices on the 3-Card Bonus bet: the gambler makes a bet, and it's resolved when the community cards are revealed. We expect that the 3-Card Bonus is not a good bet for the gambler; counting the various winning hands will allow us to determine the validity of this expectation.

- **Straight flush**: Since an ace may be either high or low in a straight flush, any card except a king can be the low card in a straight flush. Once the low card is determined, we have no choice as to the remaining 2 cards. Therefore, there are 48 possible 3-card straight flushes.

- **3 of a kind**: There are 13 ranks in a deck; once we choose one, there are $\binom{4}{3} = 4$ ways to choose the 3 cards. It follows that there are $13 \cdot 4 = 52$ three-of-a-kinds.

- **Straight**: As with straight flushes, we begin by identifying the lowest card in the straight. There are 48 choices. Once that choice has been made, the ranks of the remaining cards are determined. There are 4 choices for the suit of each of the other 2 cards, but we must subtract the number of straight flushes, so the number of straights is

$$48 \cdot 4 \cdot 4 - 48 = 48 \cdot 15 = 720.$$

- **Flush**: There are 4 ways to choose the suit, and then there are $\binom{13}{3} = 286$ ways to pick the 3 cards comprising the flush. Once again, the number of straight flushes must be subtracted, which leaves

$$4 \cdot 286 - 48 = 1096$$

flushes.

- **Pair**: We have 13 choices for the rank of the pair, and then $\binom{4}{2} = 6$ ways to pick the two cards of the pair. We then have 48 choices for the third card in the hand, so the number of pairs is

$$13 \cdot 6 \cdot 48 = 3744.$$

Adding everything up, we see that 5660 three-card hands win the 3-Card bonus bet. There are $\binom{52}{3} = 22,100$ possible 3-card hands, so the probability of a winning hand is just over 25%.

The house edge on this bet is found to be 7.28%, so while this is not a good bet in comparison with the main bet at most casino games, it ranks well in comparison with other optional side bets.

Double Down Stud

Deck composition:	52 cards
Hand size:	5 cards
Community cards:	4
Betting rounds:	3

Double Down Stud (DDS) was an attempt to adapt 5-card stud poker into both a table game and a video game. The video version of Double Down Stud was an alternative to standard video poker, in which the game is five-card draw poker; the chief differences between the two are the larger payoffs at DDS and the corresponding lower probability of winning hands due to the inability to discard and draw cards [83]. Hands are ranked in accordance with Table 4.1 (page 85).

A hand of DDS begins with each player placing a wager and receiving one card, dealt face up. The dealer or machine then deals four community cards, three face up and one face down. Players are then given the option to double their bet before the fourth community card is revealed. As in games like blackjack and video poker, player choice like this can serve to increase the casino's advantage if it is not exercised perfectly. Winning poker hands are paid off according to Table 4.17.

The probability of achieving any given hand in DDS is simply the probability of getting that hand in 5 cards. One strategy is self-evident: double your bet if you have a paying hand in four cards, as with 7♡ 5♣ J♣ J♡, before the fifth card is turned up. You're guaranteed a win, so you might as well double your payoff with absolutely no increased risk. When, though, should you double your bet (as in Mississippi Stud) in the hope that the final community card will turn your non-winning four-card hand into a winner?

TABLE 4.17: Payoff table for Double Down Stud [83, p. 136]

Hand	Payoff odds
Royal flush	2000–1
Straight flush	200–1
4 of a kind	50–1
Full house	11–1
Flush	8–1
Straight	5–1
3 of a kind	3–1
Two pairs	2–1
Pair, jacks or better	1–1
Pair, 6s through 10s	Push

Example 4.9. Suppose that you hold the $5\heartsuit$ and the first three community cards are $3\diamondsuit\ 4\clubsuit\ 6\heartsuit$, giving you a draw to an open-ended straight. Is this worth risking double your initial wager?

Whether you double your bet or not, there are eight cards—four 2s and four 7s—that will complete the straight. In addition, if the fifth card is a 6, you will push on your hand with a pair of 6s. The expectation if you do not double your bet is

$$E = (5) \cdot \frac{8}{48} + (0) \cdot \frac{3}{48} + (-1) \cdot \frac{37}{48} = \$\frac{3}{48} = \$.0625,$$

a positive value which means that you currently hold an advantage—so this hand is worth doubling. If you double the bet, the expectation also doubles, to $.125. ∎

Note that the presence of the 6 in your hand makes the difference here. If you hold an unsuited 2345, its expected value is zero, so not doubling and doubling both mean playing a fair game. If a game is fair, the only point in playing is any possible enjoyment you might receive from the excitement of the action.

Example 4.10. An inside straight, without a flush draw, may be a different matter. If you are holding $3\spadesuit\ 4\diamondsuit\ 6\spadesuit\ 7\spadesuit$, your expectation is

$$E = (5) \cdot \frac{4}{48} + (0) \cdot \frac{6}{48} + (-1) \cdot \frac{38}{48} = -\$\frac{18}{48}$$

—a negative value that only gets more negative if you double your bet. Taking back your bet is not an option, so the best strategy is not to double your bet and hope that the fifth card is one of the ten (four 5s, three 6s, and three 7s) that allow you to avoid a loss. ∎

Where is the line? It turns out that only the following non-winning four-card hands justify doubling your bet [83]:

- A four-card flush, including straight and royal flushes.

- A four-card open-ended straight, as in Example 4.9, but not an inside straight.

- An unsuited JQKA hand. This is the only closed-end straight (a straight involving an ace, which can only be completed at one end) that should be doubled.

This last hand is included because of its considerable possibility of leading to a paying pair on the fifth card. The expectation for this hand is

$$E = (5) \cdot \frac{4}{48} + (1) \cdot \frac{12}{48} + (-1) \cdot \frac{32}{48} = \$0,$$

so the double is justified since the expectation is not negative. At this point, you're playing a fair game, and as with the 2345 hand mentioned above, doubling may be the correct short-term decision as you play for a bigger immediate gain against the best odds you're likely to find in the casino.

Finding the house advantage for a single hand of DDS requires careful attention to the effects of a bet-doubling strategy. Blindly doubling every wager gives a HA of 27.4%—too much money is lost doubling hands with little or no probability of winning. At the other extreme, if you never double your bet, you face the same 27.4% HA, because you lose higher payoffs on guaranteed winning hands. The challenge in computing the theoretical house edge lies in separating out when a winning hand such as $J\Diamond\ J\spadesuit\ 3\heartsuit\ 5\spadesuit\ 9\Diamond$ arises from a doubled bet, when the facedown card is not a jack, and when it comes with a non-doubled bet, as when one of the jacks is the final card revealed.

The following hands always come from a doubled bet if the strategy detailed above is followed:

- Royal flush
- Straight flush
- Four of a kind
- Full house
- Flush

This covers 9516 of the 2,598,960 possible hands. All other final poker hands could result from doubling a wager, or not.

Example 4.11. Consider three-of-a-kind hands. There are 54,912 of these, 4224 of each rank. If the triple is of 6s through aces, at least two of them will appear in the four exposed cards, and so a player following the recommended strategy will always double his or her bet. For low triples of 2s through 5s, the hand will only be doubled if the triple occurs among the first four exposed cards; that is, if the hole card is *not* one of the triple.

The probability that a "low" three-of-a-kind hand will not be doubled is simply the chance of the hole card completing the triple: $\frac{3}{5}$—hence, $\frac{2}{5}$ of all low three-of-a-kinds will be doubled. Combining these two cases shows that the probability of a three-of-a-kind being doubled is

$$\frac{9 \cdot 4224 + 4 \cdot \frac{2}{5} \cdot 4224}{54,912} = \frac{44,774.4}{54,912} \approx .8154,$$

and so we may say that a typical winning three-of-a-kind hand carries a wager of 1.8154 betting units. ∎

Example 4.12. Two-pair hands can be analyzed similarly. There are 123,552 hands with two pairs, which can be sorted into 78 subsets of 1584 hands each based on the paired cards. These subsets can be broken down by the ranks of the pairs:

- If both pairs are "high" cards—6 through ace—then at least one such pair will be among the four known cards, and so these hands will always be doubled. There are $(9 \cdot 8)/2 = 36$ such combinations of high pairs.

- Two-pair hands where both pairs are "low" cards—2 through 5—comprise only 6 of these 78 subsets: 5-4, 5-3, 5-2, 4-3, 4-2, and 3-2. These hands will be doubled if the odd card is the hole card, an event which occurs with probability 1/5.

- "Mixed" hands, with one high and one low pair, will be doubled provided that the high pair falls among the four exposed cards, or alternately, if the hole card is the odd card or one of the low pair. There are 36 subsets of two-pair hands which contain a mixed pair, and the probability that they will be doubled is 3/5.

The total probability of two pairs being doubled is then

$$\frac{36 \cdot 1584 + 6 \cdot \frac{1}{5} \cdot 1584 + 36 \cdot \frac{3}{5} \cdot 1584}{123,552} = \frac{93,129.2}{123,552} \approx .7538,$$

giving a weight of 1.7538 betting units to the average two-pair hand. ∎

Continuing in this fashion through the remaining poker hands yields a house advantage for Double Down Stud of 2.67% [83].

Wild 52

Deck composition:	53 cards: 1 joker
Hand size:	7 cards dealt, 5 used
Community cards:	2
Betting rounds:	3

In *Wild 52*, players compete on two fronts to build the best 5-card poker hand out of 7 cards, from a deck including one wild joker. The game begins with an Ante bet from every player, who may also make two optional bets: a Bonus bet based solely on the player's best 5-card hand, and a Joker bet that pays off if their hand contains the joker. Each player receives 5 cards, and two community cards are dealt face down.

Before the first community card is turned up, players have the option of making a Play bet of twice their Ante bet, or folding and forfeiting the ante. After the reveal, players may make an Option bet, again of twice their Ante bet, or check. The second community card is then revealed and the dealer's hand announced. If the dealer qualifies with at least a pair of 5s, then any player whose hand outranks the dealer's receives even money on all of this or her bets. A player losing to the dealer's hand forfeits all wagers. If the dealer does not qualify, players' Ante bets are paid off at even money and any Play or Option bets push.

The Bonus bet pays off in accordance with Table 4.18; players can win a Bonus bet even if the dealer's hand defeats theirs or if they fold in the second round of betting. The Joker bet pays 10–1 if the joker falls as one of the two

TABLE 4.18: Wild 52: Bonus bet pay table

Hand	Payoff
Five aces	1000–1
Five of a kind: 2–K	250–1
Royal flush without joker	200–1
Royal flush with joker	150–1
Straight flush without joker	100–1
Straight flush with joker	50–1
Four of a kind without joker	25–1
Four of a kind with joker	20–1
Full house	5–1
Flush	3–1
Straight	2–1

community cards, and 4–1 if the joker is dealt to the player's hand.

Example 4.13. Three players ante $1 each for a round of Wild 52; each one also makes the Bonus and Joker bets for $1. The hands are as follows:

Player 1:	J♠	6♢	5♢	4♠	Joker
Player 2:	Q♣	10♢	10♡	9♢	8♣
Player 3:	J♢	9♣	8♡	7♠	3♠

It should be noted that the rack card for Wild 52 explicitly forbids player collaboration, so players may not share information about their hands with each other. As in Caribbean Stud Poker, player collaboration might be advantageous in deciding whether or not to make the Bonus or Option bets.

Suppose that all three players stay in the game by making the Bonus bet. The first community card is turned up: the 6♡. Player 1 has three 6s and potential to use the joker to draw a straight, Player 2 has a pair of 10s and an inside straight draw, and Player 3 has an open-ended straight draw. Players 1 and 2 make the Option bet while Player 3 checks, and the final community card is shown to be the A♡.

Player 1 holds three 6s, with $5 at risk. Player 2's hand is a pair of 10s with $5 riding, and Player 3 has an ace-high hand with $3 in play. Player 1 wins $4 on the joker bet; no player has a hand high enough to win the Bonus bet.

The dealer's cards are now revealed: 9♣ 8♠ 6♠ 4♢ 3♣. The dealer qualifies with a pair of 6s, but loses to Players 1 and 2 while beating Player 3. Player 1's net win is $8, Player 2 wins $3, and Player 3 loses $5. ∎

The Joker bet is simple to analyze. The probability of a player's hand receiving the joker is

$$\frac{\binom{52}{4}}{\binom{53}{5}} = \frac{270,275}{2,869,685} = \frac{5}{53},$$

while the probability of the joker falling to the community cards is

$$\frac{52}{\binom{53}{2}} = \frac{2}{53}.$$

The resulting expected value is

$$E = (10) \cdot \frac{2}{53} + (4) \cdot \frac{5}{53} + (-1) \cdot \frac{46}{53} = -\frac{6}{53} \approx -\$.1132,$$

so the casino holds an 11.32% advantage on the Joker bet.

If the casino paid 5–1 on a community joker, the HA would drop to 1.89%; conversely, cutting that payoff to 3–1 gives the house an edge of 20.75%. Depending on the market, any of these three payoffs might be necessary or desirable.

The 10–1 payoff for a joker in a player's hand is a convenient amount for dealers to pay off, so changing it must take ease of handling into account. Making this an 11–1 payoff while keeping the community joker odds at 4–1 would give the casino a 7.55% edge, but might slow down the game slightly, particularly if players are wagering unusual amounts on this side bet.

Wild 5 Poker

Deck composition:	53 cards: 1 joker
Hand size:	5 used out of 5 or 6
Community cards:	2
Betting rounds:	2

Wild 5 Poker is a variation on Wild 52 that streamlines betting slightly and incorporates a player choice. After placing an Ante wager and a Bonus wager of the same size, players and dealer are each dealt 5 cards, with 2 community cards dealt face down to the table. After examining their hands, each player has three options:

- Fold, forfeiting the Ante bet.
- Play the 5 cards as they stand. This requires a Play wager of 1–3 times the Ante bet.
- Discard one card and play on with access to the 2 community cards, which also requires a Play bet.

The Wild 5 Poker rack card provides the player with a simple strategy:

- **On any hand of 2 pairs or higher, play the cards as dealt and make a 3× Play bet.**

  This might be suboptimal advice. With a hand such as 3 of a kind, discarding one card for two shots at improving the hand to a full house or 4 or 5 of a kind might be a better strategy—especially since it costs nothing to discard and use the community cards.

- **If holding a pair of 8s or higher, a 4-card flush, or a 4-card outside straight, discard 1 card and make a 3× Play bet.**

- **Holding a queen or higher, or an inside straight draw, discard one card, but raise only 1×.**

- **Fold any hand which is jack-high or lower.**

As in the optimal strategy for Mississippi Stud, there is no recommendation ever that a player make a 2× Play bet, and any player who does is handing the house an additional advantage, either by underbetting a favorable hand or overbetting a marginal hand.

Meanwhile, the dealer sets his or her hand according to the house way [97]:

- **Stand on any pat five-card hand: a straight, flush, full house, straight flush, royal flush, or five of a kind.**

- **When holding two pairs, three of a kind, or four of a kind, discard the lowest singleton card.**

- **If the hand contains a 4-card flush or outside straight, discard the fifth card. If the hand contains both a 4-card flush and a 4-card outside straight, then break up the straight and keep the 4-card flush.**

- **With a pair or five unmatched cards, discard the lowest singleton. This includes the case where the dealer holds an inside straight.**

The community cards are then revealed and the hands are compared. If the dealer fails to qualify with at least a pair of 2s, Ante bets push, while Play bets pay even money if the players hand beats the dealer's, even if the dealer's hand does not qualify.

Example 4.14. When dealt

$$K\heartsuit\ Q\spadesuit\ 9\spadesuit\ 6\heartsuit\ 5\spadesuit,$$

the player should discard the 5♠ and make a 1× Play bet. If the dealer fails to qualify, a king-high hand stands a decent chance of winning the Play bet, but not so strong a chance that a 3× bet is justified. The dealer would also discard the 5♠ when faced with this initial hand.

If the community cards are then the 7♡ and 2♠, the hand would be valued as king-high. ■

Example 4.15. Suppose that the player holds

$$9\diamondsuit\ 7\heartsuit\ 5\spadesuit\ 2\spadesuit\ 2\heartsuit.$$

The correct play would be to make a 1× Play bet and discard the 5♠. If the dealer holds

$$7\spadesuit\ 5\heartsuit\ 3\spadesuit\ 3\diamondsuit\ 2\clubsuit,$$

she would discard the 2♣. Both player and dealer would then have access to the community cards.

If the community cards were the 9♡ and 5♢, the player's hand of two pair, 9s and 5s, would beat the dealer's qualifying hand of two pairs with 5s and 3s. ■

The mandatory Bonus wager pays off if the player's final hand is at least 3 of a kind and beats the dealer's hand. Its pay table is shown in Table 4.19.

TABLE 4.19: Wild 5 Poker Bonus bet pay table

Hand	Payoff
Five of a kind	100–1
Royal flush	50–1
Straight flush	10–1
Four of a kind	5–1
Full house	3–1
Flush	2–1
Straight	1–1
Three of a kind	Push

Example 4.16. A crushing blow to a Wild 5 player would be to make 5 of a kind and then lose both the Play bet and the Bonus bet when the dealer has a higher 5-of-a-kind. In poker jargon, this would be called a "bad beat." How likely is this admittedly very improbable event?

One requirement for this bad beat is that both hands have access to the joker, so it would have to be one of the community cards. Another requirement is that the player would have to disregard the strategy advice above by discarding a card and drawing to a hand containing either 3 or 4 of a kind, or a full house. (If this happens to you, you can't say you weren't warned!) The dealer's strategy calls for making this discard; that's one place where player and dealer have different strategies.

We will begin by considering the simple case that the player draws into 5 deuces, and so can be beaten by any higher 5-of-a-kind. This involves looking at 10-card subsets of the 53-card deck: 4 player cards, 2 community cards, and 4 dealer cards. There are 3 subcases to consider:

A. Player: 4 deuces.
   Community cards: Joker and a card of rank x.
   Dealer: 3 xs and 1 unmatched card.

B. Player: 4 deuces.
   Community cards: Joker and an unmatched card.
   Dealer: 4 of a kind other than x.

C. Player: 3 deuces and an unmatched card.
   Community cards: Joker and the fourth deuce.
   Dealer: Four of a kind.

Let the counts of these three cases be denoted N_A, N_B, and N_C, and denote the probability by $p(2)$. We have

$$p(2) = \frac{N_A + N_B + N_C}{\binom{53}{10}}.$$

There are 44 unmatched cards to consider. Counting player cards, community cards, and dealer cards in that order gives

$$N_A = 1 \cdot (4 \cdot 12) \cdot 44 = 2112$$
$$N_B = 1 \cdot 44 \cdot 12 = 528$$
$$N_C = \binom{4}{3} \cdot 44 \cdot 1 \cdot 12 = 2112,$$

and so

$$p(2) = \frac{2112 + 528 + 2112}{\binom{53}{10}} = \frac{4752}{19,499,099,620} \approx \frac{1}{4,103,346}.$$

In general, let n denote the rank of the player's five-of-a-kind. The formula developed above for $p(2)$ can be generalized to $p(n)$, $3 \leqslant n \leqslant 13$, by replacing every 12 in the expressions for N_A, N_B, and N_C by $(14 - n)$, the number of ranks that are higher than n. This factor counts the number of five-of-a-kinds that beat the player's hand. We have

$$p(n) = \frac{4 \cdot (14 - n) \cdot 44 + 44\,(14 - n) + 4 \cdot 44(14 - n)}{\binom{53}{10}} = \frac{396 \cdot (14 - n)}{\binom{53}{10}}.$$

With $n = 11, 12, 13$ corresponding respectively to a player hand of 5 jacks, queens, and kings, the probability of this worst of all bad beats is

$$\sum_{n=2}^{13} p(n) = \frac{30,888}{\binom{53}{10}} \approx \frac{1}{631,284},$$

or approximately the chance of being dealt a royal flush in 5 cards. ∎

Wild Hold 'Em Fold 'Em

Deck composition:	52 cards: Deuces are wild
Hand size:	5 cards
Community cards:	None
Betting rounds:	3

In *Wild Hold 'Em Fold 'Em*, 5-card stud poker players have the three-round structure of games such as Wild 52 without the need to beat a dealer's hand or other players; the only challenge to the gambler is to build the strongest possible 5-card poker hand. Five cards are dealt to each player, and all four deuces are wild, making it somewhat easier to build high-ranking hands. Part of the motivation behind this game is the desire to offer the equivalent of a live poker game without having to play against other players. There's still an element of skill, but it's skill at assessing one's hand and judging its likelihood of improving, not skill at reading other people and trying to predict their hand from their actions. Wild Hold 'Em Fold 'Em made it to the floor at Sam's Town Casino in Las Vegas.

The game begins, once again, with a mandatory Ante wager, after which each player receives 3 cards. Players may then either "Hold 'Em," continuing the hand by making a "Bet" wager equal to the Ante, or "Fold 'Em," which ends their hand and forfeits the ante. Those who continue receive a fourth card, and face a second decision: they may stay in the game by placing a "Raise" wager of twice their ante, or fold and lose both the Ante and Bet wagers. Staying in to the final card, then, requires risking four times the original Ante wager.

All remaining players receive their fifth card, which is dealt face up. Since player collaboration may decrease the house advantage, every player's first four cards are dealt face down, and players are admonished not to share hand information with each other. Winning hands are paid off according to Table 4.20. Except for a pair of aces, these odds are paid on all three bets.

TABLE 4.20: Wild Hold 'Em Fold 'Em pay table

Hand	Payoff
Natural royal flush	1000–1
Four deuces	200–1
Wild royal flush	30–1
Five of a kind	20–1
Straight flush	10–1
Four of a kind	4–1
Full house	4–1
Flush	4–1
Straight	3–1
Three of a kind	1–1
Two pairs	1–1
Pair of aces	1–1 on Ante only

It is clear that a player holding a pair of aces or three of a kind in the first 3 cards should back his or her hand through the last two bets, since a win is certain. Any hand containing a single deuce probably merits at least one more wager, since there are multiple ways for the hand to reach a pair of aces or better.

Example 4.17. If the player's first three cards are $Q\clubsuit\ 6\heartsuit\ 2\diamondsuit$, the hand will win if its last 2 cards include any deuce, ace, queen, or 6, as well as if they form a pair.

There are 13 deuces, 6s, queens, and aces remaining in the unseen portion of the deck. The probability of drawing one of them together with any card of one of the other 9 ranks is

$$\frac{13 \cdot 36}{\binom{49}{2}} = \frac{468}{1176}.$$

The deck still contains 4 cards of the 10 ranks not represented in the player's hand, and 3 cards of the represented ranks. The probability of drawing a pair, of any rank, is

$$\frac{10 \cdot \binom{4}{2} + 3 \cdot \binom{3}{2}}{\binom{49}{2}} = \frac{69}{1176},$$

making the total probability of turning $Q\clubsuit\ 6\heartsuit\ 2\diamondsuit$ into a winning hand

$$\frac{537}{1176} \approx .4566.$$

While this is less than ½, it contains the possibility of several payoffs greater than 1–1, for example, drawing a pair of queens, 6s, or deuces turns the hand into 4 of a kind. As such, even without a possible straight or flush draw, making the Bet wager is a sound move—though the hand should be reevaluated after the fourth card is dealt before making the Raise wager. ∎

A hand containing 2 or more deuces is always a winner, but Scarne's maxim ("If you have nothing, get out.") can be interpreted as a strong suggestion to fold a hand if improving it relies too heavily on drawing deuces. The decision to make the Bet and Raise wagers with a deuce-free hand that's not a guaranteed winner in the hopes that the final two cards will improve it is the challenge facing Wild Hold 'Em Fold 'Em players.

Example 4.18. If a player's first three cards are $10\diamondsuit\ 9\heartsuit\ 7\diamondsuit$, a hand with no pair, flush draw, or open-ended straight draw, folding is probably the best decision. Improving this hand to at least a pair of aces requires one of the following combinations in the remaining two cards:

- The hand wins $1 with a pair of aces if the last two cards are either two aces or an ace and a deuce. If those cards are both deuces, then the hand will be called a straight. This event has probability

$$p_1 = \frac{\binom{4}{2} + 4 \cdot 4}{\binom{49}{2}} = \frac{22}{1176}.$$

- The hand advances to two pairs or 3 of a kind if the final cards are two of the ranks represented or one such card with a deuce. The probability of this even-money win is

$$p_2 = \frac{3 \cdot \binom{3}{2} + 9 \cdot 4}{\binom{49}{2}} = \frac{45}{1176}.$$

- A straight occurs with any 8 together with any 6 or jack. A deuce can substitute for one or both of those two cards. This 3–1 payoff occurs with probability

$$p_3 = \frac{4 \cdot 8 + 4 \cdot 12 + \binom{4}{2}}{\binom{49}{2}} = \frac{86}{1176}.$$

The probability that the hand does not improve to a winning hand is then

$$p_4 = 1 - p_1 - p_2 - p_3 = \frac{1023}{1176} \approx .8699,$$

about 87%.

The expected value of $10\diamondsuit\ 9\heartsuit\ 7\diamondsuit$ if folded without making any additional bets is $-\$1$. If played through to the end, a total of \$4 is at risk, and the expectation is then

$$E = (1) \cdot p_1 + (4) \cdot p_2 + (12) \cdot p_3 + (-4) \cdot p_4 \approx -\$2.43,$$

more than twice as bad. This three-card hand should be folded.   ∎

Example 4.19. The hand $J\spadesuit\ Q\diamondsuit\ K\clubsuit$ looks somewhat stronger, but since the Wild Hold 'Em Fold 'Em pay table, unlike a standard video poker pay table, pays off only on a pair of aces rather than on a pair of jacks or better, this appearance is deceptive. The existence of an open-ended straight draw may make this hand worth backing for at least one additional card, but the numbers will tell the story. The probability of completing a straight—either 910JQK or 10JQKA—is the same $\dfrac{86}{1176}$ that we saw for $10\diamondsuit\ 9\heartsuit\ 7\diamondsuit$ in Example 4.18. The expected value improves with slightly higher probabilities for two pairs or 3 of a kind; Table 4.21 shows the number of ways to improve $J\spadesuit\ Q\diamondsuit\ K\clubsuit$ to the reachable winning hands.

TABLE 4.21: Wild Hold 'Em Fold 'Em: Ways to improve $J\spadesuit\ Q\diamondsuit\ K\clubsuit$

Hand	Number of ways
Straight	86
Three of a kind	45
Two pairs	27
Pair of aces	6

Computing the expected value of this hand when backed to the finish gives $E \approx -\$2.31$. Despite looking like a strong hand, this hand should be folded, not be backed by the Bet wager.

Suppose you do make the Bet wager, though. If the fourth card is the $7\heartsuit$, it's time to fold rather than making the Raise wager, since there's no way to draw a fifth card that will complete a winning hand. If, on the other hand, the card is the $2\diamondsuit$, the fifth card can no longer improve the hand to two pair or a pair of aces due to the wild deuce. If the fifth card is a jack, queen, or king, the hand becomes 3 of a kind. If the fifth card is an ace, 10, 9, or deuce, the hand is a straight. The chance of the fifth card improving the hand to win is

$$\frac{9+15}{48} = \frac{24}{48} = .5,$$

and the new expected value of the hand is

$$E = (4) \cdot \frac{9}{48} + (12) \cdot \frac{15}{48} + (-4) \cdot \frac{24}{48} = \frac{120}{48} = \$2.50,$$

a positive expected value that justifies the Raise wager, even as this scenario was reached by an unwise Bet wager.   ∎

As a general rule, which initial 3-card hands—aside from the obvious 3-of-a-kind or two-ace hands—justify a Bet wager? Examples 4.18 and 4.19 suggest that few hands are worth additional bets.

DJ Wild

Deck composition:	53 cards: Joker and deuces are wild
Hand size:	5 cards
Community cards:	None
Betting rounds:	2

DJ Wild (**D**euces and **J**okers Wild) is a variation on 5-card stud poker with 5 wild cards in the deck. Players make Ante and Blind bets, then receive 5 cards and must decide whether or not to challenge the dealer's hand with a Play bet of twice the Ante.

In deciding whether or not to play on in hopes of beating the dealer with a higher-ranking hand, it is necessary to assess one's hand properly in the presence of 5 wild cards. What do these cards do to the relative ranking of poker hands?

The highest-ranking hand in DJ Wild is 5 wilds, which is broken out as a hand type distinct from 5 of a kind. There is only 1 way to draw all 5 wild cards, so its probability is

$$\frac{1}{\binom{53}{5}} = \frac{1}{2,869,685}.$$

Example 4.20. Which ranks higher in DJ Wild: 5 of a kind or a royal flush?

Counting these two hands requires a decision: will a hand such as $K\diamondsuit\ 2\spadesuit\ 2\heartsuit\ 2\diamondsuit\ 2\clubsuit$, with 4 wild cards and one card from a royal flush, be counted as a royal flush or as 5 of a kind? There are 100 of these hands: 20 choices for the non-wild card and $\binom{5}{4} = 5$ ways to choose the wild cards. We shall set these "ambiguous" hands aside temporarily.

5-of-a-kinds can be counted by thinking of 5 of a kind as drawing 5 cards from a set of 9: the 4 cards of a given rank different from 2, the 4 deuces, and the joker. We need to remove the one combination consisting of all 5 wild cards and the 100 ambiguous hands described above from this collection. Since 12 ranks remain for the 5-of-a-kind, because the 2s have been ruled out, the probability of this type of 5-of-a-kind is then

$$\frac{12 \cdot \binom{9}{5} - 101}{\binom{53}{5}} = \frac{1411}{2,869,685}.$$

For royal flushes, we consider the number within a given suit. As with 5-of-a-kinds, we draw 5 cards from a set of 10, again removing from the count the 101 hands described previously. We have a probability of

$$\frac{4 \cdot \binom{10}{5} - 101}{\binom{53}{5}} = \frac{907}{2,869,685}.$$

No matter how we count the 100 ambiguous hands, a royal flush has a lower probability, and so beats 5 of a kind. ∎

The following chart shows which DJ Wild hands may and may not be made with 0–5 wild cards.

Wild Cards:	0	1	2	3	4	5
5 wilds						×
Royal flush	×	×	×	×	×	
5 of a kind	×	×	×	×	×	
Straight flush	×	×	×	×		
4 of a kind	×	×	×			
Full house	×	×				
Flush	×	×	×			
Straight	×	×	×			
3 of a kind	×	×	×			
Two pairs	×					
Pair	×	×				
High card	×					

We note that a two-pair hand cannot be made with a wild card, for a candidate hand with a single wild card must also contain a pair, and thus will be valued as three of a kind or better. Careful counting shows that in DJ Wild, two-pair hands are rarer than 3-of-a-kinds. The probability of two pairs is

$$P(\text{Two pairs}) = \frac{\binom{12}{2} \cdot \binom{4}{2}^2 \cdot 40}{\binom{53}{5}} = \frac{95,040}{\binom{53}{5}}.$$

The probability of 3 of a kind can be computed by considering three cases, based on whether the hand contains 0, 1, or 2 wild cards. Let $P(k)$ denote the probability of 3 of a kind with k wild cards. We have

$$P(0) = \frac{12 \cdot \binom{4}{3} \cdot \binom{11}{2} \cdot 4^2}{\binom{53}{5}} = \frac{63,360}{\binom{53}{5}}$$

$$P(1) = \frac{12 \cdot \binom{4}{2} \cdot 5 \cdot \binom{11}{2} \cdot 4^2}{\binom{53}{5}} = \frac{316,800}{\binom{53}{5}}$$

$$P(2) = \frac{\binom{5}{2} \cdot \left(\binom{12}{3} \cdot 4^3 - \left[\binom{12}{3} \cdot 4 - 8 \cdot \binom{4}{2} \cdot (4^3 - 4)\right]\right)}{\binom{53}{5}} = \frac{367,200}{\binom{53}{5}}.$$

In the expression for $P(2)$, the term in brackets removes possible straights and flushes from the count. We have

$$P(3 \text{ of a kind}) = P(0) + P(1) + P(2) = \frac{747,360}{\binom{53}{5}},$$

confirming the assertion that $P(\text{Two pairs}) < P(3 \text{ of a kind})$.

It is not necessary to have 5 wild cards in the deck to reverse the order of two pairs and 3 of a kind; this is also observed with a single wild card [6]. While it might be appropriate simply to reorder poker hands to place two pairs above 3 of a kind, this would just cause players to evaluate a hand with a pair and a wild card as two pairs. Some efforts to impose restrictions on the use of a wild card that preserve the relative order of two pairs and 3 of a kind have been made [89, p. 586], but they have not been widely adopted by card players. DJ Wild avoids this problem by merely pushing the Blind wager if the player's winning hand is less than a straight.

Players who beat the dealer's hand receive even money on their Ante and Play bets, and their Blind bets are paid according to Table 4.22, which reflects this hand ranking.

TABLE 4.22: DJ Wild: Blind bet pay table

Hand	Payoff odds
5 wilds	1000–1
Royal flush	50–1
5 of a kind	10–1
Straight flush	9–1
4 of a kind	4–1
Full house	3–1
Flush	2–1
Straight	1–1
Other hand	Push

Player strategy for DJ Wild is simple [92]:

Make the Play bet if your hand is at least a pair of 4s,
unless you hold a pair of 4s with a singleton 3.

For example, the hand $44AK3$ should be folded, even as you should play on with the weaker 44865. The reason for the exception is that a pair of 4s will beat a pair of 3s, and if you hold a 3, the probability that the dealer holds a pair of 3s is decreased. The probability that the dealer holds a pair of 3s is

$$\frac{\binom{4}{2} \cdot \binom{11}{3} \cdot \left(\binom{4}{1}\right)^3}{\binom{53}{5}} \approx .0221,$$

while the probability that the dealer holds a pair of 3s if the player has one 3 in her hand is

$$\frac{\binom{3}{2} \cdot \binom{11}{3} \cdot \binom{4}{1}^3}{\binom{53}{5}} \approx .0147,$$

which is smaller by a factor of $\frac{2}{3}$. Note that there is no need to consider wild cards in computing these probabilities, since any hand with a wild card must rank higher than a pair of 3s.

This means that any hand with a wild card justifies a Play bet, since a hand with a wild card (W) must be at least 8543W, which is read as a pair of 8s—better than a pair of 4s.

At the Jack Casino in Cincinnati, Ohio, DJ Wild players are forbidden from sharing information about their hands with each other. Since the dealer always qualifies, this collusion prohibition is most useful in counting the number of wild cards among the players' hands, which gives important information about the potential strength of the dealer's hand. This additional information may cause a player to play on with a comparatively weak hand or fold a stronger hand.

Consider the case of 6 players at a DJ Wild table. The probability that a player's or dealer's hand contains no wild cards is

$$\frac{\binom{48}{5}}{\binom{53}{5}} \approx .5967.$$

Since the hands dealt in a single round are not independent, the probability that no wild cards are dealt among all 6 hands cannot be computed with the Multiplication Rule. All we need to do, though, is count the number of ways to pick 30 cards without choosing any wild cards, since the distribution of the cards among the 6 players doesn't matter. We have

$$P(\text{No wilds among 6 player hands}) = \frac{\binom{48}{30}}{\binom{53}{30}} \approx .0117.$$

If 30 cards have been dealt to the players with no wild cards, the PDF for the number k of wild cards in the dealer's hand is

$$P(k) = \frac{\binom{5}{k} \cdot \binom{18}{5-k}}{\binom{23}{5}},$$

which is tabulated below.

k	$P(k)$
0	.2546
1	.4547
2	.2425
3	.0455
4	.0027
5	2.972×10^{-5}

It is most likely that the dealer holds 1 wild card; the expected number of wild cards in the dealer's hand is 1.09. Player collusion in this circumstance would most likely draw players away from making the Play bet without a better hand than the pair of 4s cited above, hence the casino forbids it.

The optional Trips bet at DJ Wild pays off based on the player's 5-card poker hand, with different payoffs (Table 4.23) for natural hands and hands using wild cards. In ranking hands, if a deuce is not used as a wild card to turn

TABLE 4.23: DJ Wild: Trips bet pay table

Hand	Payoff with wild cards	Natural payoff
5 wilds	2000–1	N/A
Royal flush	90–1	1000–1
5 of a kind	70–1	70–1
Straight flush	25–1	200–1
4 of a kind	6–1	60–1
Full house	5–1	30–1
Flush	4–1	25–1
Straight	3–1	20–1
3 of a kind	1–1	6–1

the hand into a winning 3-of-a-kind hand, the hand is considered natural. For example, the hand 9♠ 9♢ 9♣ 4♢ 2♠ is read as a natural 3-of-a-kind paying 6–1, not a wild full house paying 5–1. Calling the hand a wild 4-of-a-kind would result in the same 6–1 payoff.

Example 4.21. A natural full house pays 30–1, while a full house with a wild card pays 5–1. Is a natural full house 6 times less likely than a wild full house?

The probability of a full house with no wild cards is

$$p_1 = \frac{12 \cdot \binom{4}{3} \cdot 11 \cdot \binom{4}{2}}{\binom{53}{5}} = \frac{3168}{2,869,685}.$$

The choices for the ranks are 12 for the triple and 11 for the pair since we may not choose 2s into either position.

The only wild full house consists of two pairs and a wild card, since a hand with a pair and more than 1 wild card will be read as four of a kind or better. A wild full house has probability

$$p_2 = \frac{\binom{12}{2} \cdot \binom{4}{2}^2 \cdot 5}{2,869,685} = \frac{11,880}{2,869,685}$$

and so we have $p_2/p_1 = 3.75$, not 6.   ■

Lunar Poker

Deck composition:	52 cards
Hand size:	5 or 6 cards
Community cards:	None
Betting rounds:	2–3

At its core, Caribbean Stud Poker is a game of 5-card stud poker played between the player and the dealer. In a similar vein, *Lunar Poker* is effectively a player vs. dealer draw poker game. The opportunity to discard and draw cards clearly gives the players a better game. How do the rules of Lunar Poker maintain the casino's advantage?

One answer might be "In part, by being complicated and challenging to follow, with several opportunities for players to make suboptimal choices." Lunar Poker is regarded as one of the most complicated carnival games ever to reach a casino floor [50]. Play begins with each player making an Ante wager and a Super wager, which must equal or exceed the Ante. The Super wager is a required bet that pays off on the player's initial 5 cards as in Table 4.24. Players may also make this wager on another player's hand, or on the

TABLE 4.24: Lunar Poker: Super bet pay table

Hand	Payoff odds
Four of a kind	200–1
Five picture cards	150–1
Full house	100–1
Flush	60–1
Straight	30–1
Three of a kind	8–1
Ace-king-queen	5–1
Five cards, same color	2–1

dealer's hand.

After the Ante and Super bets are placed, players and the dealer are then each dealt 5 cards; one dealer card is dealt face up. Super bets on player hands are then resolved; Super bets on the dealer's hands wait until that hand is revealed before being paid. Players then have the option of either folding or paying the amount of their Ante bet to do one of the following:

- Exchange 2–5 cards from their hand for new cards. Note that a 1-card discard is not permitted.

- Buy a 6th card. The player's hand must then be chosen using the best 5 cards from 6.

The duplicate Ante bet here is a fee paid to the house; it is immediately collected and removed from play, and receives no betting action.

Buying a 6th card is likely to be most attractive to players holding an initial hand within 1 card of a high hand and its corresponding high payoff. One reason for allowing players to exchange cards might be that players holding weak hands may be inclined to pay the additional amount to draw new cards in hopes of improving their hand rather than folding, thus giving the casino a chance at collecting more money.

Example 4.22. With an initial no-pair hand of $K\diamondsuit\ 3\heartsuit\ 10\diamondsuit\ 5\spadesuit\ 8\clubsuit$, Lunar Poker strategy suggests folding if the dealer's upcard is a 5 or higher, and exchanging 4 cards (all but the king) otherwise. A player drawing to the king against a high dealer card is offering the casino a greater advantage. ∎

Players then stay in the game by making a Bet wager, which must be double their ante, or fold. A player may play out his or her initial 5-card hand from this point. The dealer does not have the options of exchanging or adding a card, and qualifies with at least Ace-King high. Player and dealer hands are then compared, with the higher hand winning. Bet wagers on winning hands are paid in accordance with Table 4.25, while Ante bets push. If the dealer

TABLE 4.25: Lunar Poker: Bet pay table

Hand	Payoff odds
Royal flush	100–1
Straight flush	50–1
4 of a kind	20–1
Full house	7–1
Flush	5–1
Straight	4–1
3 of a kind	3–1
2 pairs	2–1
Pair	1–1
Ace-King	1–1

fails to qualify, the Ante bets of players still in the hand are paid at even money if their hands are less than a straight; a straight or higher pays 4–1 on the Ante bet in this circumstance.

Lunar Poker also offers a Double Combination Payout: If a player can make two different 2–5 card poker hands of Ace-King or higher, at least one of which beats the dealer's hand, then he or she is paid from Table 4.25 on both hands. The hands must not overlap completely; each must contain at least one card that the other does not. This payoff eliminates the need for players to discard a single card and replace it by drawing.

Example 4.23. If the player's cards are $8\clubsuit\ 7\clubsuit\ 6\clubsuit\ 10\diamondsuit\ 9\spadesuit$ and she buys a sixth card, the $J\spadesuit$, then her two straights (J10987 and 109876) both pay 4–1, provided that the dealer's hand is lower than a jack-high straight.

The Double Combination Payout can also be won by a 5-card hand. The hand $9\diamondsuit\ 5\heartsuit\ 9\clubsuit\ K\diamondsuit\ A\clubsuit$ can be read either as a pair of 9s or as an Ace-King high hand. For the purposes of the Double Combination Payout, only the two 9s are used to construct the pair of 9s hand, and the AK hand consists of AK5 and one of the two 9s. This fulfills the requirement that there be at least one card in each combination that is not used in the other one. If the dealer's hand qualifies but cannot beat a pair of 9s, the player wins even money twice. ∎

Gaming writer and mathematician Michael Shackleford, the "Wizard of Odds," computed that there are 627,392,769,491,403,000,000 possible outcomes to a Lunar Poker game, a number which confounds computer analysis [93]. Some work from the game's owner has produced a basic strategy for players [51]:

- **Buy a 6th card when holding any straight or flush.**
- **Play 4 of a kind without buying a 6th card.**
- **On a full house, buy a 6th card if your cards do not match the dealer's upcard; play on otherwise.**

In these three cases, the player holds a strong hand that is likely to win. The goal in buying a sixth card with a strong hand is to trigger the Double Combination Payout by generating a second paying hand. In the case of 4 of a kind, there is very limited opportunity to improve— spending an Ante wager in search of a pair has negative expectation.

Example 4.24. Suppose, though, that you hold four 9s and a king, and the dealer's upcard is the 4♠. Buying a 6th card might lead to an Ace-King hand as well as a pair of kings. Is this possibility worth buying a 6th card?

There are 7 cards, 4 aces and 3 kings, that lead to a second paying hand. Since both AK hands and pairs pay the Bet hand at even money, you would be spending $1 to win $2. The expected value of this additional dollar is

$$E = (2) \cdot \frac{7}{46} + (-1) \cdot \frac{39}{46} = -\frac{25}{46},$$

a negative value which advises against buying the 6th card.

Simply make the Bet wager, with the high probability that you will turn $2 into $42. ■

- **If you hold 3 of a kind, exchange 2 cards if the dealer's upcard is a different rank, play if it is.**

This is a simple quest to turn 3 of a kind (3–1 payoff) into 4 of a kind (20–1). Without that possibility, the additional chance of drawing into a full house does not justify the expense of the sixth card. If the dealer is seen to hold that fourth card, of course, there's no point in chasing it.

- **Play two pairs without further action.**

Since full houses only pay 7–1, the return on buying a 6th card again falls short of its cost, as with 3 of a kind.

- **Holding one pair, buy a 6th card with any 4-card straight or flush, as, for example, with**

$$A\spadesuit \ A\clubsuit \ K\spadesuit \ 6\spadesuit \ 4\spadesuit.$$

- **Play when holding a pair at least as high as the dealer's upcard and exchange 3 cards if your pair is lower than the upcard.**

  This recommendation is one part offense—the pursuit of the 5-card straight or flush—and one part defense. A pair is not a sure winner, even though the dealer cannot exchange cards, and the second sentence here is advice to improve the pair when necessary. If the dealer holds one card higher than your pair, the chance of a higher pair rises, and you can't beat that—hence the advice to exchange 3 cards.

- **With an Ace-King hand, buy a 6th card if your hand contains an outside straight, flush, or inside 6-card straight. If not, exchange 3 cards.**

  For example, $A\clubsuit$ $K\heartsuit$ $J\clubsuit$ $10\diamondsuit$ $9\spadesuit$ can be completed to a 6-card straight with any queen, and so it's worth buying a 6th card. If holding $A\clubsuit$ $K\heartsuit$ $10\diamondsuit$ $9\spadesuit 8\heartsuit$, you should discard and replace all but the ace and king.

  If the hand is lower than AK, two possible strategies are given:

- **Buy a 6th card with an outside straight, a 4-card flush, or a 6-card inside straight draw. Otherwise, exchange all cards except for aces and face cards.**

  A hand with no cards higher than 10 should be folded; this is another case where sound strategy recognizes the right time to cut your losses and get out.

- **Exchange 4 cards, holding a queen or higher, if the dealer's upcard is a 5 or lower, as in Example 4.22. Fold if these conditions are not met.** This advice foreshadows the Bet or Fold decision that will arise after cards are bought or exchanged in its advice to hold queens, kings, or aces. If the dealer's upcard is small, the chance of a qualifying dealer hand drops to a point where the risk of exchanging 4 cards is worth taking.

Once the decision to exchange cards or buy a 6th card is made, the following strategy is suggested for a player's final hand.

- **Make the $2\times$ Bet wager if holding AKQ or higher.**

- **Fold anything less.**

The beacon hand for Lunar Poker is simply AKQ32, the lowest AKQ hand.

The complexity of this strategy, and of the game, may cause players to adopt a simpler strategy of their own choice. This plays into the casino's hands, as any deviation from this optimal method of play will raise the casino's edge at Lunar Poker from the 4.90% that this set of choices provides [93].

We will consider some simple questions about Lunar Poker.

Example 4.25. Find the probability of being dealt 5 picture cards, which pays 150–1 on the Super bet. This hand is called a *blaze* in some local versions of poker.

The probability of drawing 5 picture cards from the 12 in a deck is simply

$$\frac{\binom{12}{5}}{\binom{52}{5}} = \frac{792}{2,598,960} \approx \frac{1}{3282}.$$

Of course, not every hand with 5 pictures pays 150–1; if the hand contains all 4 queens and the $J\heartsuit$, it's 4 of a kind and qualifies for a 200–1 payoff. There are 24 ways to draw 4 of a kind where the 5th card is also a face card, so the true number of 5-picture-card hands in Lunar Poker is 768. ∎

Table 4.24 for the Super bet stops at 4 of a kind, because Lunar Poker offers players a "free" bonus bet at the outset of a new hand in the Instant Payout. This bet pays 1000–1 on a player's Ante if their first 5 cards form a royal flush and 250–1 if those cards form a straight flush, while immediately ending the hand with no further action. (The royal flush and straight flush payoffs in Table 4.25 apply only if the player has exchanged cards or purchased a sixth card.) If the hand is anything lower, the Ante bet is not lost and the game proceeds as usual, so Instant Payout carries no risk for the gambler. How much does this game feature cost the casino?

The expected value of this bonus is

$$E = (1000) \cdot \frac{1}{649,740} + (250) \cdot \frac{9}{649,740};$$

there is no negative term because the player does not lose his or her Ante bet on an initial hand that's less than a straight flush. We have

$$E = \frac{3250}{649,740} = \frac{25}{4998} \approx \$.0050$$

—half a cent, meaning that the casino is effectively paying ½% of the cumulative Ante wagers back to players with Instant Payout.

Of course, Instant Payout isn't truly "free": the small advantage that Lunar Poker players gain from this rare payout simply balances a casino advantage elsewhere in the game.

Single-Wager Variations

The poker games discussed in this subsection include only one round of betting and function like roulette by calling for gamblers to wager on the composition of a single hand common to all players.

Poker-All

Poker-All was a variation on keno incorporating playing cards and poker hands that debuted in 1979 at the Sahara Casino in Reno. The game used 52 ping-pong balls in a blower, each one labeled with a different playing card. The casino drew 6 of the 52 balls each game, and players were offered a variety of wagers on the individual balls and on various poker hands that could be made with the best 5 of 6 cards.

One source of the casino's advantage in Poker-All came from the structure of several bets that centered on certain cards being high (9 through ace) or low (2 through 7) [72]. The 8s were classified as neither high nor low, and so the house won all high and low bets when an 8 was drawn. For example, players could make separate even-money High or Low bets on each of the 6 balls. The probability of winning either bet was $\frac{6}{13}$, and the probability of losing was $\frac{7}{13}$, which yielded a 7.69% HA.

We note that Poker-All could easily have been played with a deck of cards dealt to the table; the balls were not essential to the game's design. Keno, as played in most casinos and state lotteries, is easily shown to have a high house edge on most bets, often well over 20% [7]. Many Poker-All wagers were no exception—and this may be an argument in favor of using ping-pong balls: players might expect better wagers from table games such as blackjack and baccarat, that are played with actual cards.

Example 4.26. Another pair of high/low bets could be made on the final 5-card poker hand. If 5 of the 6 balls were either high or low, as selected by the bettor, the bet paid 12–1.

Since the bets are mathematically equivalent, we focus on the Low bet for convenience. One simple way to assess this bet would be to break the event "Win" into 2 mutually exclusive events with 5 and 6 low balls drawn. This gives the probability p of winning as

$$p = P(5 \text{ low balls}) + P(6 \text{ low balls})$$
$$= \frac{\binom{24}{5} \cdot 28}{\binom{52}{6}} + \frac{\binom{24}{6}}{\binom{52}{6}} = \frac{1,324,708}{20,358,520} \approx .0651.$$

The expected value of either wager is then

$$E = (12) \cdot p + (-1) \cdot (1 - p) = 13p - 1 \approx -\$.1541.$$

■

Example 4.27. The *Any Ace* bet paid even money if at least 1 ace was drawn. Using the Complement Rule, we find that the probability of a win was

$$1 - P(\text{No aces}) = 1 - \frac{\binom{48}{6}}{\binom{52}{6}} \approx .3972.$$

The same chance of winning applied to the *Any 8* bet. Both bets carried a HA of 20.55%. ■

Any 8 was accompanied by 4 separate bets that the 8 of a specified suit would appear among the 6 balls, each of which paid 8–1. On its face, this appears to offer eight times the payoff of Any 8 while keying on a result that looks to be only one-fourth as likely, so it ought to have about half the HA, right? Let's see: The probability of any given playing card, whether an 8 or not, showing up in the 6-ball draw is

$$\frac{\binom{51}{5}}{\binom{52}{6}} = \frac{3}{26},$$

so the expected value of a $1 bet is

$$(8) \cdot \frac{3}{26} + (-1) \cdot \frac{23}{26} = \frac{1}{26}$$

—a rare positive expectation, and a player advantage of 3.85%.

This unexpected edge comes from reading the Poker-All layout's payoff language of "1 wins 8" as an 8–1 payout rather than 8 for 1 [72]. If the payoff is 8 for 1, then the casino retrieves the advantage, which is then 7.69%.

We move next to bets on specific poker hands. The *Flush* bet paid 80–1 if at least 5 of the 6 balls were of the same suit. The specific suit does not matter, so we will look at the probability of a diamond flush, and we can compute the probability of winning by considering two mutually exclusive events.

$$P(\text{Diamond flush}) = P(5 \diamondsuit s) + P(6 \diamondsuit s)$$
$$= \frac{\binom{13}{5} \cdot 39}{\binom{52}{6}} + \frac{\binom{13}{6}}{\binom{52}{6}} = \frac{48,477}{20,358,520} \approx .0024.$$

Multiplying this by 4 to cover all suits gives $P(\text{Win}) \approx .0095$. The HA is 22.85%: once again high for card games, but in the common range for keno.

A suit bet that was easier to win paid 28–1 if 4 or more of the 6 balls were of the suit chosen by the gambler. A similar bet paid 6–1 if any 4 balls were of the same suit, with no need for the player to make a choice. If we let p be the probability of winning the first bet, the expectation of the first bet is

$$E = (28) \cdot p + (-1) \cdot (1 - p) = 29p - 1,$$

while the expectation of the second is

$$E = (6) \cdot 4p + (-1) \cdot (1 - 4p) = 28p - 1,$$

which is slightly lower. Calculation of p is left as Exercise 10.

The Poker-All *Full House* bet paid 100–1 if the 6 balls included a full house. With 6 cards to choose from, there are 3 ways to draw a full house:

- Three of a kind, a pair, and a sixth card matching neither group.
- Two 3-of-a-kinds.
- Four of a kind and a pair.

The chance of drawing a full house then increases from about 1 in 694 to

$$\frac{\left[13 \cdot \binom{4}{2} \cdot 12 \cdot \binom{4}{2} \cdot 44\right] + \left[\binom{13}{2} \cdot \binom{4}{3}^2\right] + \left[13 \cdot 12 \cdot \binom{4}{2}\right]}{\binom{52}{6}},$$

where the numerator counts the three hands above in order. This fraction evaluates to .0082, approximately 1 chance in 122. A sixth card makes drawing a full house about 5.5 times more likely.

The casino claims an edge of 17.19% on this wager.

Pokerbo

Deck composition:	312 cards, 6 decks
Hand size:	5 cards
Community cards:	5

The Crown Casino in Melbourne, Australia hosts a New Games Lab on its gaming floor, where new casino carnival games can be tested in a live casino environment. Among the games offered in the Lab is *Pokerbo*, another table game based on 5-card poker. This simple game calls for players to make bets on the composition of a 5-card hand which is dealt to the table. There is no discarding and drawing of cards; Pokerbo in this sense is a simulation of a single hand of 5-card stud poker. One major difference between the two games is that Pokerbo is dealt from a 6-deck shoe, which makes 5 of a kind a possible hand and changes the probabilities of other hands.

Three types of Pokerbo bets are available:

- Four **Single outcome** wagers may be made on the lowest-ranking five-card hands. A bet on No Hand pays 1–1 if a high-card hand is dealt. The Pair bet pays 1–1 on a pair, Two Pair pays 13–1, and Three of a Kind pays 20–1.

- Higher poker hands may be bet as a unit with the **Straight 'n' Up** wager. This bet is riskier, since the dealt hand must be at least a straight to win, but offers higher payoff odds. The pay table is shown in Table 4.26.

- **Specific Suit** wagers may be made on any of the four suits, and pay off according to Table 4.27 if 2 or more cards are of that suit.

  It is worth noting that at least one suit must appear on 2 or more cards in any hand; thus at least one Specific Suit wager will win on any Pokerbo hand. It is possible for 2 Specific Suit wagers to pay off on the same 5-card hand.

TABLE 4.26: Pokerbo: Straight 'n' Up pay table

Poker hand	Payoff odds
Royal flush	500–1
Straight flush	400–1
Five of a kind	300–1
Four of a kind	150–1
Full house	100–1
Flush	50–1
Straight	25–1

TABLE 4.27: Pokerbo: Specific Suit pay table

Cards in suit	Payoff odds
2 cards	1–1
3 cards	2–1
4 cards	5–1
5 cards	50–1

Example 4.28. If the Pokerbo hand is

$$A\spadesuit\ 7\diamondsuit\ 2\clubsuit\ 5\heartsuit\ 5\diamondsuit,$$

then the Pair bet pays 1–1 and a Specific Suit bet on diamonds pays 1–1. All other bets lose.

The hand

$$A\heartsuit\ 5\heartsuit\ A\diamondsuit\ 9\diamondsuit\ 5\clubsuit$$

pays off three ways:

- The Two Pairs bet pays 13–1.
- A Specific Suit bet on either hearts or diamonds pays even money. ∎

The probability of a royal flush in a 1-deck game is 1 in 649,740. In Pokerbo, with its 6-deck shoe, the probability drops slightly, to

$$\frac{4 \cdot 6^5}{\binom{312}{5}} \approx \frac{1}{766,988}.$$

By contrast, the probability of 5 of a kind rises from 0 in a single-deck game to

$$\frac{13 \cdot \binom{6}{5}}{\binom{312}{5}} = \frac{1}{305,851,084},$$

which suggests that 5 of a kind deserves a higher payoff on the Straight 'n' Up bet than a royal flush.

Example 4.29. Suppose that a gambler simultaneously bets $1 each on all 4 suits, which guarantees at least 1 winning bet. Construct the PDF for the outcome of the combined wagers.

The net payoff depends on how the 4 suits are distributed among 5 cards. Table 4.28 shows the possible distributions with their net payoffs.

TABLE 4.28: Pokerbo: Distribution of suits and corresponding net profit on 4 simultaneous $1 suit bets

Suit distribution	Net profit
2–1–1–1	–$2
2–2–1–0	$0
3–1–1–0	–$1
3–2–0–0	$1
4–1–0–0	$2
5–0–0–0	$47

If the distribution is $a - b - c - d$, then the number of ways to select the cards, assuming that the hand is dealt from a fresh shoe, is

$$A \cdot \binom{78}{a} \cdot \binom{78}{b} \cdot \binom{78}{c} \cdot \binom{78}{d},$$

where A is a factor that counts the number of ways to assign the suits to the quantities a, b, c, and d. For example, in the 2–2–1–0 case, there are $\binom{4}{2} = 6$ ways to pick the two suits with 2 cards each and 2 ways to pick the suit of the singleton card, so $A = 12$. The probability of a hand with suits distributed 2–2–1–0 is then

$$\frac{12 \cdot \binom{78}{2} \cdot \binom{78}{2} \cdot \binom{78}{1} \cdot \binom{78}{0}}{\binom{312}{5}} = \frac{8,440,856,424}{23,856,384,552} \approx .3538.$$

Values for A and the corresponding probabilities are shown in Table 4.29.

TABLE 4.29: Pokerbo: Probability of suit distribution in 5-card poker hands

Suit distribution	A	Probability
2–1–1–1	4	.2389
2–2–1–0	12	.3538
3–1–1–0	12	.2328
3–2–0–0	12	.1149
4–1–0–0	12	.0560
5–0–0–0	4	.0035

Combining Tables 4.28 and 4.29 gives the PDF for a player's winnings, Table 4.30.

TABLE 4.30: PDF for 4 simultaneous $1 suit bets at Pokerbo

Suit distribution	Net profit	Probability
2–1–1–1	–$2	.2389
2–2–1–0	$0	.3538
3–1–1–0	–$1	.2328
3–2–0–0	$1	.1149
4–1–0–0	$2	.0560
5–0–0–0	$47	.0035

From this table, we can draw the following conclusions:

- The probability of making a profit on this $4 bet is approximately .1744, slightly better than 1 chance in 6.

- The expected return on a $4 wager is –31.7¢, and the house advantage on this collection of bets is 7.94%.

∎

This example presumes a fresh 6-deck shoe. As hands are dealt from the shoe, the relative proportion of the 4 suits is likely to change, which will affect the probabilities in Table 4.30. However, unless one suit is unusually absent or present, this bet is not likely ever to have a positive player edge. If one suit is known to dominate in the undealt portion of the shoe, a single Specific Suit bet on that suit may be a better choice.

4.2   Combination Card Games

New casino games that simply combine elements from other games, which we shall call *combination games*, are frequently unsuccessful. In the words of game designer Dan Lubin, such games tend to be like mixing gin and Kahlua—both may be fine beverages on their own, but combining them produces something that is rather less enjoyable than either one is individually [50].

That doesn't stop game designers from trying, and occasionally a combination card game reaches the casino floor.

Royal 89

Royal 89 is a game from eTable Games which adds a poker element to baccarat. A round of Royal 89 begins with two cards, from a single deck, dealt to each player and to the dealer—these will be played out and counted similar to baccarat hands. Two major game play differences between baccarat and Royal

89 are that each player is dealt his or her own hand and competes individually against the dealer, and that all hands receive a third card regardless of the total of the first two. Three bets are available to the player: the required main Score bet and two optional wagers: the Royal Bonus bet which must equal the Score bet, and a Royal Combo bet which pays off based on the six cards in the player and dealer hands.

The fact that a third card is mandatory might, in play, break up a dealt natural 8 or 9. Royal 89 insures the player against this possibility by offering a "Free Money" payoff that pays off the Score bet at 1–1 if the player's first two cards total 8 or 9. This payoff is made independent of the dealer's hand, which is not exposed until all player hands have been dealt.

The Score bet pays off as shown in Table 4.31. This bet pushes, however,

TABLE 4.31: Royal 89: Score bet pay table

Player hand	Payoff
3 of a kind	25–1
One pair	1–1
Higher than dealer hand	1–1

if the dealer's hand is 0.

The two Royal bets are based on face cards. The Royal Bonus bet pays based on the player's completed 3-card hand, as in Table 4.32:

TABLE 4.32: Royal 89: Royal Bonus bet pay table

Player hand	Payoff
3 Kings	60–1
3 Queens or 3 Jacks	30–1
3 face cards	5–1
2 face cards	2–1
1 King or Queen	Push

Finally, the Royal Combo bet is paid based on the number of face cards in the player and dealer hands combined (Table 4.33):

TABLE 4.33: Royal 89: Royal Combo bet pay table

Face cards	Payoff
6	200–1
5	30–1
4	5–1
3	3–1

The player may withdraw a Royal Combo bet if neither of his or her first two cards are face cards; these cards from ace through 10 are called "score cards."

Example 4.30. In a two-handed game against the dealer, suppose that each player has made a $1 bet on all three spots and that the following cards are dealt:

> Player 1: 6♢ 2♠
> Player 2: J♠ 2♣

Player 1's total is 8, and so he collects $1 from the Free Money payoff. His Score bet remains in play. Since both of his cards are score cards, he pulls back his Royal Combo bet, which can at this point win at most $5.

Player 2's total is 2. Since she holds a face card, her Royal Combo bet must stay on the board.

The dealer's cards remain face down. A third card is dealt to each hand:

> Player 1: 9♣. Total = 7.
> Player 2: A♣. Total = 3.

Neither player has a winning Royal Bonus hand, and so each loses $1 on that bet. The dealer's cards are now exposed:

> Dealer: K♡ 6♣ 3♣. Total = 9.

The dealer's hand beats both players' hands, and so their Score bets are lost. Neither player has three of a kind or a pair, so neither qualifies for the bonus. Player 2 also loses her Royal Combo bet, as her hand and the dealer's contain only two face cards between them. Player 1 has a net loss of $1; Player 2 lost $3.   ∎

Example 4.31. The only other place where a strategy may be developed for Royal 89 is in the choice to withdraw the Royal Combo bet; every other action is strictly determined by the rules of the game.

Assume that you're playing Royal 89 head-to-head against the dealer. If you are dealt two score cards, should you pull back your Royal Combo bet, as did Player 1 in Example 4.30?

By withdrawing this bet, your expected value is $0. You should let the bet ride if its expectation—given that you have two non-face cards—is positive. At this point, the bet only pays off if the four cards yet to be revealed—your third card and the dealer's hand—contain at least three face cards. This seems unlikely.

$$P(3 \text{ face cards}) = \frac{\binom{12}{3} \cdot \binom{38}{1}}{\binom{50}{4}} = \frac{8360}{230,300} \approx .0363,$$

and

$$P(\text{4 face cards}) = \frac{\binom{12}{4}}{\binom{50}{4}} = \frac{495}{230,300} \approx .00021.$$

The probability of losing this bet is $\frac{221,445}{230,300}$—over 95%—and the expectation is

$$E = (5) \cdot \frac{495}{230,300} + (1) \cdot \frac{8360}{230,300} + (-1) \cdot \frac{221,445}{230,300} = -\frac{210,610}{230,300} \approx -\$.9145.$$

It is rare for a casino to offer a player a chance to back out of a losing proposition, so such an opportunity should be seized whenever possible. You would be wise to withdraw this bet if you are not dealt any face cards in your first two cards. ∎

Triple Shot

One ambitious combination game that saw actual casino play was *Triple Shot*, a single-deck card game combining Casino War, blackjack and poker. Six cards were dealt to each player, and several to the dealer:

- The first card dealt to each side was used in a hand of Casino War. Going to war was not an option; tied hands lost half the War wager.

- The War cards were then treated as the first cards of blackjack hands. As many cards as necessary were dealt to each hand, subject to the modified blackjack rules in use:

  1. Dealer hits soft 17.
  2. Naturals pay 6–5.
  3. Double down on any 2 cards. This rule is partial compensation for rules 1 and 2, which increase the casino's edge.
  4. Only aces may be split. Only one card is dealt to split aces, as is standard in blackjack.
  5. A six-card total under 21 automatically wins.
     Rules 4 and 5 are designed to limit a player to no more than 6 cards in the blackjack phase of the game, so as to eliminate the possibility of having to remove cards from play when the game moves on to poker.

  The HA of single-deck blackjack with these rules was 1.84% [96].

- Additional cards were then dealt as needed to round each player's hand out to 6 cards, and their best 5-card poker hands were paid off according to Table 4.34. The dealer did not receive a poker hand, and so there was no showdown or question of the dealer qualifying.

- The dealer's blackjack hand was then dealt out to completion.

TABLE 4.34: Triple Shot: Poker pay table [96]

Poker hand	Payoff odds
Royal flush	100–1
Straight flush	30–1
Four of a kind	15–1
Full house	7–1
Flush	5–1
Straight	4–1
Three of a kind	3–1
Two pairs	2–1
Pair, queens or better	1–1

Separate wagers, which did not have to be of equal amounts, were required on each of the games selected by the gambler—who did not have to play all 3 games.

Example 4.32. Suppose that all 3 bets are made and that the first cards dealt are 9$\diamond$ to the player and 6$\clubsuit$ to the dealer. The player wins on the Casino War bet.

The next card is dealt to the player to complete her initial blackjack hand, and it's the 5$\spadesuit$. Single-deck basic strategy calls for standing on a 14 against a dealer 6, so she does.

The game moves on to the poker hand. The next 4 cards dealt to the player are 8$\clubsuit$ $Q\diamond$ 5$\heartsuit$ 2$\spadesuit$. The best 5-card hand here is a pair of 5s, which does not qualify for a poker payout.

With the poker bet resolved, we return to the dealer's blackjack hand. He draws the 2$\clubsuit$ followed by the 6$\diamond$ and $A\heartsuit$ before busting with the $J\diamond$. The player then wins her blackjack bet. ∎

For the Casino War bet, the guaranteed 50% loss when the initial two cards tie replaces a war, where the expected loss for a single-deck game when going to war is 32.1% (page 95). This 17.9% shift in the casino's favor factors into the HA for the War bet. For this wager, the probability of tying the dealer is $\frac{3}{51}$. The probability of winning the War bet depends on the player's card, from 0 when dealt a 2 to .9412 on an ace. Averaging over all 13 equally likely possible ranks gives $P(\text{Win}) = P(\text{Lose}) = \frac{24}{51}$. The resulting expectation is

$$E = (1) \cdot \frac{24}{51} + (-1) \cdot \frac{24}{51} + \left(-\frac{1}{2}\right) \cdot \frac{3}{51} = -\frac{1}{34},$$

and we see that the casino's edge arises entirely from the treatment of ties, as wins and losses cancel each other out. The HA on this bet is 2.71%.

Computing the HA on the poker portion of a Triple Shot bet is a simple matter of finding the probabilities of achieving each paying 5-card hand in 6 cards. There are 4 5-card royal flushes, each of which may be combined with any of the other 47 cards, which gives $4 \cdot 47 = 188$ 6-card royal flushes. The probability of a royal flush is then

$$\frac{188}{\binom{52}{6}} = \frac{188}{20,358,520} \approx 9.234 \times 10^{-6},$$

or once every 108,290 hands.

Example 4.33. The probability of four of a kind is also easy to compute. Given one of 13 four-of-a-kinds, there are then $\binom{48}{2} = 1128$ ways to pick the other 2 cards, making the probability of four of a kind

$$\frac{13 \cdot 1128}{\binom{52}{6}} = \frac{14,664}{\binom{52}{6}} \approx 7.203 \times 10^{-4}.$$

∎

The remaining probabilities are shown in Table 4.35. Combining these

TABLE 4.35: Triple Shot: Poker probabilities [96]

Poker hand	Probability
Royal flush	9.234×10^{-6}
Straight flush	8.311×10^{-5}
Four of a kind	7.203×10^{-4}
Full house	.0082
Flush	.0101
Straight	.0178
Three of a kind	.0360
Two pairs	.1244
Pair, queens or better	.1106
Nonpaying hand	.6922

probabilities with the payoffs in Table 4.34 gives a poker HA of 3.20%.

Given the HAs of 2.71% for War, 1.84% on blackjack, and 3.20% on poker, the best betting strategy for Triple Shot would be to wager only on blackjack, the game with the smallest HA. A player interested in following this optimal strategy can do better by finding a conventional blackjack table offering better rules.

Show Pai

Show Pai debuted in 2016 at the Palace Station casino in Las Vegas. This game is a combination game that merges Pai Gow poker play and baccarat scoring. Players and dealer are each dealt four cards from a six- or eight-deck shoe and must arrange them into a High hand of three cards and a Low hand with one card. Hands are summed and scored according to baccarat counting rules: aces count 1, 2s through 9s count their face value, and 10s and face cards count 0. Any tens digit in the resulting sum is discarded, so hands all have scores from 0 through 9. The High hand must have a value at least that of the Low hand, or the hand is invalid. Invalid hands are called *No Pai*.

Example 4.34. If dealt

$$K\Diamond\ 8\Diamond\ 10\spadesuit\ 8\heartsuit,$$

the best play is to put $K\Diamond\ 8\Diamond\ 10\spadesuit$ in the High hand and $8\heartsuit$ in the Low hand, making two hands of 8. ∎

Example 4.35. One possible No Pai hand is

$$4\spadesuit\ 4\heartsuit\ 4\Diamond\ 4\clubsuit,$$

which must be set with the High hand totaling 2 and the Low hand valued at 4. ∎

The player's two hands are compared to the dealer's, with the hand closer to 9 winning as in baccarat. The player wins even money if he or she wins both hands or wins one and pushes the other, unless the dealer has No Pai. In this circumstance, the player pushes if he or she can set two valid hands and loses if also holding No Pai. If each side wins one hand, the round is a push. If the dealer and player make two hands that both push, the player loses.

Example 4.36. Suppose that the player is dealt $6\heartsuit\ 3\spadesuit\ 2\heartsuit\ 4\clubsuit$ and the dealer holds $9\heartsuit\ 4\Diamond\ 2\clubsuit\ 5\Diamond$.

- The player would set $3\spadesuit\ 2\heartsuit\ 4\clubsuit$ as her High hand, totaling 9, and place the $6\heartsuit$ as the Low hand with a value of 6.

- The dealer's hand could not place the $9\heartsuit$ as the Low hand, because this would leave a lower High hand of 1. This hand could be played as 8 and 2, 6 and 4, or 5 and 5. Since the house way for setting Show Pai hands calls for maximizing the Low hand, the dealer would play a High hand of $9\heartsuit\ 4\Diamond\ 2\clubsuit$, totaling 5, and a Low hand of the $5\Diamond$, also a 5.

The Player wins both hands, and thus collects on her Play bet. ∎

One scenario where the casino gains an advantage over the players is on a dealer No Pai, where the best a player can do is push even though the dealer has no valid hand. This is asymmetrical, because the player loses when holding No Pai. Another source of the house edge at Show Pai is the player loss when both hands push the dealer's.

These bad beats are among the several game possibilities covered by the optional Bonus bet, which functions as a kind of insurance against several player-unfriendly game outcomes and provides a bonus for some exceptionally good player hands: 9–9 and 8–8. One pay table for the Bonus bet is shown in Table 4.36.

TABLE 4.36: Show Pai Bonus bet pay table

Result	Payoff odds
Double No Pai	80–1
Player has all 10s	25–1
Either hand has No Pai	10–1
Player has 9–9	5–1
Player has 8–8	4–1
Double push	2–1

If the player's High and Low hands both push the dealer's hands, a player backing up a $1 Play bet with a $1 Bonus bet would then lose the Play bet but win 2–1 on the Bonus bet. The net result would be a win of $1.

Example 4.37. What is the probability of a Player hand consisting of four 10s, which must be played with the High and Low hands both totaling 0, but which pays 25–1 if the Bonus bet is made?

If Show Pai is dealt from a 6-deck shoe, there are ninety-six 10s among the 312 cards in the shoe. Assuming a fresh shoe, the probability of receiving four 10s is

$$\frac{\binom{96}{4}}{\binom{312}{4}} = \frac{3,321,960}{387,278,970} \approx .0086 \approx \frac{1}{117}.$$

∎

While this is not the only outcome leading to a winning Bonus bet, it is one that seems vulnerable to card counting. An excess of 10-count cards could improve this probability to a point where the bet had a positive player expectation.

Flip-It

Flip-It combined elements from poker and roulette into a new table game that debuted in a field trial at the Rio Casino in Las Vegas in 2018. An eight-deck shoe with 3 jokers per deck was used, for a total of 440 cards. These jokers were not used as wild cards; their effect in the game was to win all bets for the house.

Four cards were dealt face up as a prelude to betting, and wagers could then be made on the fifth card out of the shoe. The layout for a Flip-It bettor looked complicated, comprising 19 different betting circles, but the available wagers could be sorted into three categories [43]:

- 12 circles are available for *Call-it* bets on the rank of the next card; every rank from 2 through king is available (ace and joker are omitted). These bets pay 10–1 on a successful prediction.

- 6 circles offer *Inside* bets: roulette-like wagers on red/black, high (8-king)/low (2–7), and odd/even. Queens (12) are even, while jacks (11) and kings (13) are odd. If the 5th card is an ace, only the red and black bets have action; all others push. If the fifth card is a joker, all Inside bets lose. Winning bets pay even money.

- The final circle is for the *Bonus* bet, which pays off on the poker hand formed by the five cards, provided that the fifth card is used to form the winning combination. This bet may only be made if the first 4 cards contain at least a pair or 4 cards to a royal flush. The pay table is shown in Table 4.37; note that straights, flushes, and straight flushes do not pay the Bonus bet. The Bonus bet is not available if the first 4 cards are merely 4 to a flush, straight, or straight flush.

TABLE 4.37: Flip-It: Bonus bet pay table [43]

Hand	Payoff
Royal flush	50–1
5 of a kind	14–1
4 of a kind	7–1
3 of a kind	6–1
Full house	5–1
Two pairs	2–1

Example 4.38. If the first four cards are $A\spadesuit K\diamondsuit 10\heartsuit 10\heartsuit$, then any 10, king, or ace will trigger a winning Bonus bet with three of a kind or two pairs. ∎

Once five cards are dealt and the bets resolved, the first card dealt is removed from the table unless the fifth card is a joker. A joker on the fifth card is removed from play. This leaves 4 active cards. A new round of betting then begins on the next card from the shoe.

Example 4.39. In Example 4.38, suppose that the fifth card was the $K\clubsuit$, forming two pairs. For the next hand, the $A\spadesuit$ would be removed and a fifth card drawn to $K\diamondsuit 10\heartsuit 10\heartsuit K\clubsuit$. Though this hand contains two pairs, the Bonus bet will only pay off if the next card is another 10 or king, forming a full house, as the fifth card must be a part of the winning combination and not an extra card.

Alternately, one could say that the Bonus bet only wins if the fifth card moves the hand upward on the pay table from where the four-card hand lies. ∎

The house advantage at Flip-It derives from the aces and jokers. The expected value on a Call-it bet from a fresh shoe depends on how many cards of the selected rank are on the table. In general, this bet should not be made if the card you choose is on the board. If four cards have been dealt and none are of the chosen rank, the expectation is

$$E = (10) \cdot \frac{32}{436} + (-1) \cdot \frac{404}{436} = -\frac{84}{436} \approx -\$.1927,$$

corresponding to a HA near 20%.

In more generality, if the board shows x cards of the chosen rank, where $0 \leqslant x \leqslant 4$, we have

$$E = (10) \cdot \frac{32 - x}{436} + (-1) \cdot \frac{404 + x}{436} = -\frac{84 + 9x}{436}.$$

As we expect, this decreases with increasing x: if you make a Call-it bet on 3 with four 3s showing, the HA rises to 27.52%.

If the 24 jokers are removed from the shoe, the expectation on Call-it rises to

$$E = (10) \cdot \frac{32}{412} + (-1) \cdot \frac{280}{412} = -\frac{40}{412} \approx -\$.0971,$$

and the HA falls to 9.71%—which, while high, is more in line with carnival game HAs.

Part of the appeal of Call-it bets to game designers and operators is that a 10–1 payoff is easy to administer at the tables: two red $5 chips for every white or blue $1 chip. Nonetheless, raising the payoff to 12–1 while retaining the jokers would produce a reasonable HA of 4.59% if no cards of the chosen rank have appeared.

Example 4.40. For the even-money bets other than Red and Black, there are 56 cards, the aces and jokers, that win both bets for the house. Again, the return on this bet depends on the cards that appear, but if we assume 2 red and 2 black cards among the first 4, we have

$$E = (1) \cdot \frac{190}{436} + (-1) \cdot \frac{246}{436} = -\frac{56}{436} \approx -.1284,$$

and the HA remains high: 12.84%. ∎

Example 4.41. The Red and Black bets do not push on a 5th-card ace. Consider the bet on Red at the top of a fresh shoe. Define r to be the number of red cards among the first 4 dealt and j to be the number of jokers among the first 4 cards. After the first 4 cards are dealt, the deck contains $208 - r$ red cards, $204 + r + j$ black cards, and $24 - j$ jokers, so $208 - r$ cards will win for Red and $228 + r$ will lose. The expectation is

$$E = (1) \cdot \frac{208 - r}{436} + (-1) \cdot \frac{228 + r}{436} = \frac{-20 - 2r}{436}.$$

This is negative for all r; the best-case scenario arises when $r = 0$. Since the four cards are dealt to the table before this bet needs to be made, a bettor can minimize this negative return by only betting if the first 4 cards were all either black or jokers. The resulting HA is 4.59%, which is substantially less than the HA on the other even-money bets considered in Example 4.40. ■

Card counting seems like it would be a big weapon in Flip-It; with so many bets, it's not unthinkable that one or more might rise to a positive expectation once, say, 3 decks have been dealt out. The challenge is that a simple blackjack High-Low count would not give enough information. For maximum effect, it would be necessary to track all of the card ranks and make the appropriate Call-It bet when the edge turned to the player. Similar counting could also be used against the Inside bets, and possibly on the Bonus bet.

Example 4.42. If, for example, 200 of the 440 cards have been dealt without the $A\Diamond$ appearing at all, and the board shows $10JQK\Diamond$, the chance of the fifth card completing a royal flush has risen from $\frac{1}{55}$ to $\frac{8}{240} = \frac{1}{30}$. The expectation of the Bonus bet is then

$$E = (50) \cdot \frac{1}{30} + (-1) \cdot \frac{29}{30} = \$.70 > 0,$$

so the player holds a 70% advantage.

What is the chance of this deck occurring? The probability of dealing 200 cards with no aces of diamonds appearing is

$$\frac{\binom{432}{200}}{\binom{440}{200}} = \frac{432! \cdot 240!}{440! \cdot 232!} = \frac{240 \cdot 239 \cdot \ldots \cdot 233}{440 \cdot 439 \cdot \ldots \cdot 433} \approx .0074,$$

approximately once in every 135 shoes. ■

There are other favorable situations worth tracking, but counting all of them at the same time would seem to be too arduous for practical use under casino conditions.

Example 4.43. Let's turn this around. Consider the Call-It bet on 6. How many cards need to be dealt from the shoe with no 6s appearing before the bet has positive expectation?

Let x be the number of cards dealt. We seek to solve the equation

$$E = (10) \cdot \frac{32}{440 - x} + (-1) \cdot \frac{408 - x}{440 - x} = 0,$$

or

$$\frac{320 - 408 + x}{440 - x} = \frac{x - 88}{440 - x} = 0.$$

If $x > 88$, slightly more than 1.6 joker-enhanced decks, then $E > 0$ and the player has the edge on a Call-It bet.

This, of course, applies to any rank from 2 through king. ■

4.3 Matching Games

Some carnival card games pay off when certain cards match in suit, rank, or color.

Pell

Pell was a short-lived card game devised by a University of North Carolina student, Talton Pell [68]. It had a brief field trial in 1982 at Sam's Town. The game was played with two separate six-deck shoes; one card was dealt from each shoe for each round. Players were offered a collection of ten wagers on the suits, ranks, and colors of the two dealt cards; the layout is shown in Figure 4.1. Each player faced one of these triangles, and could bet on any or all of the options.

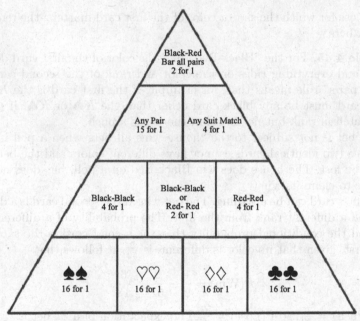

FIGURE 4.1: Player options for Pell [68].

Quick calculations will show that, as written, all but one of these bets (the Black-Red bet at the top) are fair: they pay off at true odds and there is no house advantage. The casino derived its edge in Pell from the further rule that if the two cards were identical—as, for example, two 7♠s—then the house won all bets. This identical pair was called a *pell*.

Example 4.44. What is the probability that the two cards match, giving all bets to the house?

The rank and suit of the first card do not matter; we are only interested in the chance that the second card matches it. Pell is a game where the cards dealt at each turn are not independent of past results—a fact exploited by card counters during the game's trial at Sam's Town—but if we start from two fresh six-deck shoes, the probability that the second card matches the first is a constant $\dfrac{1}{52}$. ∎

The value that we have computed here is a long-term average; the exact value at any deal depends on the cards that remain in the two shoes.

The ten Pell wagers may be divided into two types:

- Those for which the suit, color, or value of the first card does not matter. These are the Black-Red bet at the top of the layout, Any Pair and Any Suit Match in the second row, and the "Black-Black or Red-Red" bet in the middle of the third row from the top.

- Those for which the suit or color of the first card matters: the remaining six bets.

Example 4.45. For the "Black-Red" bet, the color of the first card does not matter, and everything rides on the color and rank of the second card. The "bar all pairs" rule means that, for example, if the first card is the $K\heartsuit$, the second card must be any black card other than the $K\spadesuit$ or $K\clubsuit$. If the two cards match in rank but not in color, the bet is a push.

This bet is not subject to the "house wins all bets when a pell is dealt" rule, since two identical cards cannot have different colors and the bet would already be lost. The house does win Black-Red on a pell, but does not need that rule to claim its win.

The first card can be anything. The bet wins if the second card is a different color *and* a different rank from the first. The probability of a different color is ½, and the conditional probability that the second card is the same rank as the first, given that its color is different, is $\frac{2}{52}$. It follows that

$$P(\text{Win}) = \frac{1}{2} - \frac{2}{52} = \frac{6}{13},$$

and $P(\text{Push}) = \frac{2}{52}$, so $P(\text{Lose}) = \frac{1}{2}$. The expectation of a $1 bet is

$$E = (1) \cdot \frac{6}{13} + (0) \cdot \frac{2}{52} + (-1) \cdot \frac{1}{2} = -\$\frac{1}{26},$$

approximately −$.0385. ∎

Example 4.46. The four suit bets in the last row of the layout are mathematically identical, and fall into the class of Pell bets where the first card matters. What is their house advantage?

These four bets win if both cards dealt are of the selected suit, *provided* that the ranks do not also match. For convenience, let $p^* = \frac{1}{52}$, the probability of choosing two matching cards computed in Example 4.44.

Once again, we assume two full six-deck shoes. The chance that the first card is the selected suit is $\frac{1}{4}$. The probability that the second card creates a winning pair, by being of the selected suit but a different rank than the first card, is $\frac{1}{4} - p^*$. The probability that the second card does not create a winning pair is then $\frac{3}{4} + p^*$. Noting that the Pell layout states that the payoff is "16 for 1," which means that the payoff includes the initial \$1 wager, the expectation on a \$1 bet is

$$E = (15) \cdot \left(\frac{1}{4}\right) \cdot \left(\frac{1}{4} - p^*\right) + (-1) \cdot \left[\frac{3}{4} + \frac{1}{4} \cdot \left(\frac{3}{4} + p^*\right)\right]$$

$$= -4p^* = -\frac{1}{13} \approx -\$.0768.$$

The HA is approximately 7.68%.

∎

It should be noted that the minimum bet at Pell when it was played was \$2; this does not, of course, change the HA when stated as a percentage. The calculations above confirm that if $p^* = 0$, corresponding to a game without the "identical pairs are a house win" rule, the expectation is also zero, and thus this bet is fair.

Example 4.47. The Pell "Any Pair" bet is definitely affected by the house rules on pells. As with Black-Red, the suit and rank of the first card do not matter; we then want the second card to have the same rank, but a different suit.

The probability of winning an Any Pair bet is then

$$\frac{1}{13} - p^* = \frac{3}{52}$$

and the probability of losing is $\frac{49}{52}$. The expectation of a \$1 bet is

$$E = (14) \cdot \frac{3}{52} + (-1) \cdot \frac{49}{52} = -\frac{7}{52} \approx -\$.1346,$$

corresponding to a 13.46% house advantage and tagging Any Pair as a bet to be avoided.

∎

Example 4.48. As with the suit bets in Example 4.46, the Black-Black and Red-Red bets are mathematically identical. Beginning with two full shoes, their expectation is

$$E = 4 \cdot \left(\frac{1}{2}\right) \cdot \left(\frac{1}{2} - p^*\right) - 1 = -\frac{1}{26} = 2p^*,$$

and so the house edge is 3.85%.

∎

Note that as long as no pells are dealt, 9 out of the 10 bets on the Pell layout are fair. This point was made by game promoters during the game's trial period: a player, it was said, could conceivably play Pell for hours making fair bets without ever bucking a house advantage. How likely was that claim?

If we use the infinite deck approximation, then the probability of a pell on any one hand is $\frac{1}{52}$, so the probability of not getting a pell in n consecutive hands is

$$\left(\frac{51}{52}\right)^n.$$

At a relaxed pace with several players, a Pell dealer could probably deal and pay off a hand in about a minute, giving 60 hands per hour. In 1 hour, the probability of seeing no pells is about .3119, so a pell-free hour would not be an unusual occurrence. The probability of no pells in 2 hours ($n = 120$) is just under 10%, so the promoters' claim is plausible. However, if we solve the equation

$$\left(\frac{51}{52}\right)^n = .50$$

for n, we get $n \approx 35.7$, meaning that in 36 hands (or 36 minutes), we expect a 50% chance of at least 1 pell.

Throughout this analysis, we have assumed two fresh six-deck shoes in order to keep the model workable. It is easy to see that Pell may be attacked by counting cards: for example, if the first 26 cards out of each shoe are red (an admittedly unlikely occurrence, with probability 2.2900×10^{-17}), then black cards exceed red cards 156 to 130 in the cards to be played, and thus Black-Black is likely to be a favorable wager. With this depleted deck, we have

$$P(\text{Black-Black}) = \left(\frac{156}{286}\right)^2 \approx .2975.$$

Since no black cards have yet been dealt, the probability of a black pell (for Black-Black loses on a red pell even without the pell rule) is

$$p = \left(\frac{1}{26}\right)^2,$$

and the expectation on Black-Black is

$$E = 4 \cdot \left[\left(\frac{156}{286}\right)^2 - \left(\frac{1}{26}\right)^2\right] - 1 \approx .1842,$$

for a player edge of 18.42%.

While this is an extreme example, it also comes with a very high player advantage, and so it is not unreasonable to think that smaller imbalances between the number of red and black cards, if detected, would still lead to exploitable betting situations. Similarly, tracking the number of cards of various suits as they are dealt might reveal imbalances among the suits that would turn one or more of the four bottom-row bets to the player's advantage.

The possibility of this approach was uncovered during the game's trial run [31], although Talton Pell himself was not convinced that the mathematics used to construct the counting strategy was valid. Indeed, while Pell readily admitted that card counting might be a workable option for players, he also held that the house advantage in his game was "entirely flexible and random"—which is not the kind of endorsement that leads to multiple casino adoptions for a new game [68]. Depending on the count in use, card-counting Pell targets either the Black-Black or Red-Red bets, or—in a more advanced counting scheme—the suit bets. For the color bets, all that is necessary is to track an imbalance of one color over the other, since one may bet on either Red-Red or Black-Black according to which color is favored.

A recommended card-counting strategy for Pell, which won its originators several thousand dollars per day in a brief assault on the game at Sam's Town, considers the makeup of each pair of cards dealt from the shoes [31]. Beginning with a count of 0, the running count is kept by adding 1 each time a Red-Red combination is dealt, subtracting 1 for each Black-Black combination, and adding 0 when the two cards dealt are of different colors. Since only two cards are dealt during each round of Pell, keeping an accurate count is considerably easier than in blackjack. Betting strategy at Pell is based solely on the sign of the running count—a positive RC favors the Black-Black bet, and a negative RC favors Red-Red. By contrast with blackjack, the advantage indicated by the count at Pell is always with the player provided he or she can keep the count and make the appropriate bet, since a negative count just signals for a Red-Red bet rather than indicating a deck composition favoring the casino.

Example 4.49. We shall track the running count through several hands of Pell:

First card	Second card	RC
$A\heartsuit$	$Q\diamondsuit$	+1
$8\heartsuit$	$10\clubsuit$	+1
$10\clubsuit$	$3\spadesuit$	0
$6\diamondsuit$	$8\diamondsuit$	+1
$3\diamondsuit$	$J\heartsuit$	+2
$5\heartsuit$	$2\spadesuit$	+2
$K\heartsuit$	$J\diamondsuit$	+3

At a running count of +3, the Black-Black bet is now favored. ■

As in blackjack, the running count is converted to a true count by dividing by the number of decks remaining in either shoe. The true count is used in Pell simply to quantify the player edge. A true count greater than 1 or less than −1 indicates that the player has an edge on the appropriate wager; this edge is computed by multiplying the TC by 4% and then subtracting 4% (which represents the approximate house advantage on the Red-Red or Black-Black bets, as found in Example 4.48). It may be easily seen that whenever the absolute value of the TC is greater than the number of decks remaining, the gambler making the proper bet is playing with an advantage [31].

A more advanced card-counting system for Pell targets the suit bets and aims to identify an imbalance of the suits in both shoes. As with blackjack card-counting systems that involve side counts of aces, the counting method described above can be expanded to track one or more suits, though trying to count all four suits would probably be best handled by a team of collaborators. In addition to maintaining the standard red/black count, a side count of the chosen suit is tracked separately. For each hand of Pell, count as follows [31]:

- +1 if your suit is not represented among the two cards.

- −1 if your suit appears once.

- −3 if both cards are of your suit.

This count is balanced, in the sense that the sum of all four suit counts after any hand will be 0.

Example 4.50. Consider the cards dealt in Example 4.49, with four side counts added for the suits.

First card	Second card	RC	♠	♡	♦	♣
$A\heartsuit$	$Q\diamondsuit$	+1	+1	−1	−1	+1
$8\heartsuit$	$10\clubsuit$	+1	+2	−2	0	0
$10\clubsuit$	$3\spadesuit$	0	+1	−1	+1	−1
$6\diamondsuit$	$8\diamondsuit$	+1	+2	0	−2	0
$3\diamondsuit$	$J\heartsuit$	+2	+3	−1	−3	+1
$5\heartsuit$	$2\spadesuit$	+2	+2	−2	−2	+2
$K\heartsuit$	$J\diamondsuit$	+3	+3	−3	−3	+3

As Black-Black was slightly favored in Example 4.49, it is not surprising to see that both spades and clubs have a positive count here. ■

For maximum betting advantage, the running suit counts should be converted as above to true counts by dividing by the number of decks remaining. Since each count is tied to a specific suit, the corresponding suit bet only carries a player edge if its true count is positive. Each swing of ±1 in the true count corresponds to the same change of 4% in the house edge that the red/black count carries, and since this bet has a base HA of 7.68% (Example 4.46), the edge swings to the player only when the true count is at least +2. An advantage to the card counter at Pell is that one is not required to make the same bet on every hand, and so opting against the ♦♦ bet until the diamond count turns in your favor is a workable strategy that should attract little attention from casino surveillance personnel.

Sam's Town did fight back against this counting scheme, by cutting off four decks from the six in each shoe and reshuffling after two decks had been dealt out, and by barring confirmed card counters, but the word was out that Pell was beatable. With Pell so ripe for attack by card counters, its lifetime on the casino floor was destined to be short. Gambling scholar Arnold Snyder raised the possibility that Pell was intentionally designed to draw card counters and assist casinos in identifying them; Talton Pell refused to comment on this idea [103]. What is known is that during the game's trial run at Sam's Town, players who were barred from Pell for counting cards also found themselves barred from the blackjack tables—whether they had been playing blackjack or not. After its month-long trial period, the game has not been seen, and it is not currently listed on the Nevada Gaming Commission's list of approved casino games [64].

Top Rung

Top Rung is a more recent game from Ubetcha Games that has some of the elements of Pell in that the suits and colors of dealt cards determine the outcome, and ranks do not matter. The game is named for the "ladder" that a player climbs with a sequence of successive bets, and the dealt cards are called *ladder* cards.

The player makes an initial bet on one or more of the four suits and two colors, trying to predict what the next ladder card will be. If this bet wins, it is paid off and the initial wager is moved "up the ladder," where the same set of suit or color choices is available and the payoff odds increase. The player may switch to a different suit or color than that of the initial bet. The ladder has four levels, with payoffs shown in Table 4.38.

TABLE 4.38: Top Rung Payoff Odds

Card #	Suit Bet Odds	Color Bet Odds
1	2–1	0–1
2	4–1	1–1
3	6–1	2–1
4	10–1	5–1

An additional Top Rung rule provides for double the listed payoff if the ladder card is an ace.

The first win on a Color bet does indeed pay off at zero to 1; the chance to advance to the next level is the only reward. For a bet that is 50/50 at the start, and would have a positive expectation on any payoff of even money or higher *if* a player is tracking the cards as they're dealt and betting on any color which has been dealt less to that point, this is a necessary feature in maintaining an edge for the casino.

Example 4.51. Suppose that four players each make a $1 bet on one of the 4 suits and the opposite color, as a form of insurance. When the first card is the 10♡, the player who bet on hearts and black wins $1 on his suit bet while losing the color bet, and the players who bet on either black suit and red see their suit bets lose and their color bets rise to the next level, though there is no payoff. The player betting on diamonds and black loses both her bets.

If the next card is the 6◇, the two red bettors win even money on their second win while the hearts bet loses at the second level. If the bets are left up on red, a third straight red card will result in a 2–1 payoff and a chance for 5–1 on the color of the fourth card. ■

The game is played with a four-deck shoe; it is soon apparent to a careful observer that successive cards are not independent and thus that mathematical analysis of this game will be challenging. To simplify the mathematics here, we use the infinite deck approximation; this will give us a reasonably accurate approximation to the true house edge. We consider a bet on the suit of the ladder card, and begin with the expectation of the *last* bet. The probability distribution for the payoff of the last bet is

Payoff	10	20	–1
Probability	$\dfrac{12}{52}$	$\dfrac{1}{52}$	$\dfrac{3}{4}$

and the corresponding expectation of a $1 bet, denoted here by X, is

$$X = (10) \cdot \left(\frac{12}{52}\right) + (20) \cdot \left(\frac{1}{52}\right) + (-1) \cdot \left(\frac{3}{4}\right) = \frac{101}{52}$$

—slightly less than $2.

This assumes that the gambler has successfully predicted three straight suits previously and has "earned" access to this positive-outcome wager. This is done by winning the third bet in the sequence. The probability distribution for this bet, including the expectation of the fourth bet, is

Payoff	$6 + X$	$12 + X$	–1
Probability	$\dfrac{12}{52}$	$\dfrac{1}{52}$	$\dfrac{3}{4}$

Let the expectation of this bet be Y. We have

$$Y = (6 + X) \cdot \left(\frac{12}{52}\right) + (12 + X) \cdot \left(\frac{1}{52}\right) + (-1) \cdot \left(\frac{3}{4}\right) = \frac{45}{52} + \frac{X}{4} = \frac{281}{208}.$$

We can continue analyzing the Suit bet path in this way. If Z is the expected return on the second bet, then

$$Z = (4 + Y) \cdot \left(\frac{12}{52}\right) + (8 + Y) \cdot \left(\frac{1}{52}\right) + (-1) \cdot \left(\frac{3}{4}\right) = \frac{17}{52} + \frac{Y}{4} = \frac{553}{832}$$

—still positive.

Finally, the expectation for a Suit bet is the following:

$$E = (2 + Z) \cdot \left(\frac{12}{52}\right) + (4 + Z) \cdot \left(\frac{1}{52}\right) + (-1) \cdot \left(\frac{3}{4}\right) = -\frac{151}{3328} \approx -\$.0454,$$

for a house edge—under the infinite-deck assumption—of about 4.54%.

Ubetcha Games' advertising lists a Suit bet HA of about 4.2%; this approximation is in close accord with that value [110].

Top Rung can be beaten by a card-counter who is careful to track the suits or colors of cards as they appear and wager on those colors or suits that are most favored. Advantage player James Grosjean, author of the book *Beyond Counting*, developed a workable card counting strategy for Top Rung that led to the game's removal from an Indiana casino [46].

Racing Card Derby

Racing Card Derby is an electronic table game that simulates a four-horse race with a single standard deck of cards. The suits of successively dealt cards correspond to the running horses; as each card is dealt, the horse of that suit moves one "length" forward on the racetrack. The winning horse is the first whose suit is dealt four times. After the winning horse is identified, the deal continues to identify the place and show horses. The first card not of the winning suit selects the second-place horse, and the next suit's horse takes third.

Bets may be made on the winning horse's color or suit, or on the top two (quinella) or three (Exact Trifecta) horses in order. The base payoff on a winning color is 1–1, and a winning suit bet pays 3–1. A quinella pays off at 10–1, and the trifecta bet pays 1–1 if only the winning suit is correct, 2–1 if the first two horses are chosen correctly, and 10–1 if all three match [77].

As with Pell, a quick calculation will show that, so far, the color and suit bets have an expectation of 0; these are fair bets with no house advantage.

Example 4.52. Symmetry suggests that each color has probability ½ of winning, and each suit has winning probability ¼. The expected value of a suit bet is then

$$E = (3) \cdot \frac{1}{4} + (-1) \cdot \frac{3}{4} = 0,$$

and is similar for the color bet. ∎

There is no reason for any one suit to be favored or disfavored, and no way—yet—for the casino to profit from Racing Card Derby.

The casino gains its edge from the 2♠, which is called the "Protest Card" [77]. If the 2♠ is dealt during the first stage of the race, before the winning horse is determined, then payoffs on the winning color drop to 1–2 and a winning suit bet pays only 2–1. This mimics the idea of an inquiry changing the outcome of a live horse race. There is no change to the payoffs if the 2♠ is dealt after the race is won and the place and show horses are being determined.

Example 4.53. Consider Table 4.39, which shows a sequence of dealt cards and the running counts of each suit.

TABLE 4.39: Sample Racing Card Derby game

Card	♣	♢	♡	♠
K♡	0	0	1	0
4♡	0	0	2	0
3♠	0	0	2	1
A♣	1	0	2	1
K♣	2	0	2	1
10♢	2	1	2	1
3♣	3	1	2	1
2♠	3	1	2	2
(Payoffs reduced)				
8♣	4	1	2	2
(Clubs wins)				
J♠	4	1	2	3
(Spades places)				
7♣	5	1	2	3
(No decision)				
K♢	5	2	2	3
(Diamonds shows)				

The race ends with the winning trifecta ♣♠♢. A color bet on black pays 1–2, and a suit bet on clubs pays 2–1, because the protest card appeared. ■

The quinella and trifecta wagers here are unaffected by the protest card. Since the order of finish matters, the four suits lead to $_4P_2 = 12$ different possibilities for the win and place horse, and so the expected value of a \$1 quinella bet is

$$E = (10) \cdot \frac{1}{12} + (-1) \cdot \frac{11}{12} = -\$\frac{1}{12} \approx -\$.0833.$$

For the trifecta, there are $_4P_3 = 24$ possible ways to arrange the top three finishers, and the expectation is

$$E = (10) \cdot \frac{1}{24} + (2) \cdot \frac{1}{24} + (1) \cdot \frac{4}{24} + (-1) \cdot \frac{18}{24} = -\$\frac{2}{24} \approx -\$.0833$$

—the same average return.

Calculating the HA on the suit bets is slightly more involved due to the protest card—but one thing that is clear is that a winning bet on spades is more likely than a bet on the other suits to be reduced to a 2–1 payoff, because the 2♠ advances the spades horse while simultaneously cutting the payoff odds. A similar reduction applies to a color bet on black.

High Card Flush

High Card Flush debuted at the 2012 Cutting Edge Table Games Conference, where it was named Best New Game [44]. The idea behind High Card Flush is simple: to draw the largest flush in a 7-card hand dealt from a single deck. Players compete head-to-head against the dealer. Each player makes an Ante bet before the deal, and once the cards are dealt, may fold their hand or make a Raise bet if they believe that their hand contains a higher flush than the dealer's. The maximum Raise bet is based on the length of the player's best flush, which is called the *maximum flush*: 2- to 4-card flushes can bet the amount of the ante as a Raise bet, 5-card flushes can be backed by twice the Ante, and 6- or 7-card flushes can bet up to three times the Ante. Good hands thus receive a reward in the form of the opportunity for a bigger Raise bet precisely when the chance of that bet winning is higher.

Example 4.54. The hand

$$A♡\ 9♣\ J♣\ 2♣\ K♣\ 10♡\ 3♠$$

contains a 4-card maximum flush, in clubs. ∎

As in CSP and similar games, the dealer must qualify for Raise bets to have any action. If the dealer holds at least a 3-card 9-high flush, then the player's and dealer's hands are compared. Flushes of equal length are compared by looking at the highest cards, so ♠$K87$ beats ♣$Q95$, although it would lose to any longer flush, even one consisting entirely of lower cards such as ◇6432. It's possible for two flushes to tie if the component cards have the same rank, as ◇$A862$ and ♡$A862$; suits are not ordered in this game. If the player's hand beats the dealer's, both the Ante and Raise bets are paid even money. If the dealer fails to qualify, the Ante bets pay even money, but all Raise bets push.

Example 4.55. Suppose that the player has been dealt

$$10♡\ A◇\ 5◇\ K♡\ A♠\ 3♡\ K◇.$$

The maximum flush in this hand is ◇$AK5$, which beats the ♡$K103$. If the dealer's hand is

$$Q♠\ J◇\ 2◇\ 4♡\ 3♣\ K♣\ 8◇,$$

then the dealer qualifies with ◇$J82$, but the player's hand wins. ∎

Example 4.56. What is the probability that a dealt 7-card hand will have no flush of length greater than 2?

The 7 cards must be distributed 2–2–2–1 among the 4 suits. There are 4 choices for the suit with a singleton card, so the probability is

$$\frac{4 \cdot \binom{13}{1} \cdot \binom{13}{2}^3}{\binom{52}{7}} \approx .1845,$$

slightly less than 1 chance in 5. ∎

At the other end of the luck spectrum, the probability of a 7-card flush is

$$P(\text{7-card flush}) = \frac{4 \cdot \binom{13}{7}}{\binom{52}{7}} \approx \frac{1}{19,491}.$$

While this certainly merits a 3× Raise bet, a thoughtful gambler realizes that if he or she holds a 7-card flush, the probability of the dealer also holding one increases. If 7 cards of one suit are removed from the deck, the dealer's 7-card flush must necessarily be of a different suit. The new probability is

$$P(\text{Dealer has 7-card flush} \mid \text{Player has 7-card flush}) = \frac{3 \cdot \binom{13}{7}}{\binom{45}{7}} = \frac{1}{8815},$$

which, while still very small, is more than double the initial probability.

This (admittedly extreme) example points up the potential for player collusion at High Card Flush, as knowledge of some or all of the players' cards might give useful information about the strength of the dealer's hand. At a full table of 6 players, complete sharing of information would reveal 42 of the cards in the deck, leaving only 10 possible cards for the dealer's hand. This could make a profound difference in player decisions—if, for example, the 10 cards were known to contain no more than 3 of any one suit and no cards above an 8, the dealer cannot qualify, and everyone should make the Raise bet, even if holding a weak hand. A Monte Carlo analysis of this level of player collaboration showed that players could conceivably achieve a 7.3% edge over the casino, but that perfect knowledge of the cards was necessary, and that fewer than 6 collaborating players was not enough to turn the game in the players' favor [42].

A simple casino countermeasure against this threat would be to allow no more than 5 players at the table at a time, which could be implemented in the design phase by directing the equipment manufacturer to provide felt table layouts with only 5 betting circles.

3 Card Blitz

3 Card Blitz took the format of High Card Flush—seven dealt cards per hand—and changed the rules from "build the biggest flush" to "play the highest 3-card flush" with a variation on blackjack scoring used to determine both the winner and the pay structure of a second bet.

Play begins with equal Ante and Blind bets from each player. An optional Flush bonus bet is also available. From 7-card hands, players and dealer form their best suited hand using up to 3 cards, with aces counting 11, face cards 10, and all other cards their point value. If a player wishes to fold at this point, he or she may do so and forfeit both bets. Players continuing on make a Play bet equal to their Ante and then face off against the dealer, with the higher hand winning. The highest possible hand is a Blitz: a 31 consisting of an ace and two 10-count cards. A Royal Blitz consists of a suited ace, king, and queen.

Winning Ante and Play bets are paid at even money. If the player wins, the Blind bet pushes unless the player's hand scores 27 or more points. In that case, Table 4.40 shows the payoffs. Double Blitz pays 50–1 if both player and dealer have a Blitz.

TABLE 4.40: 3 Card Blitz: Blind bet pay table

Hand	Payoff
Double Blitz	50–1
Royal Blitz	8–1
31 Blitz	4–1
27–30	1–1
26 or less	Push

Example 4.57. If a player is dealt

$$J\spadesuit\ 3\heartsuit\ 2\heartsuit\ A\diamondsuit\ 8\clubsuit\ 7\clubsuit\ 5\clubsuit,$$

then the best possible hand is 20, consisting of the $8\clubsuit\ 7\clubsuit\ 5\clubsuit$. ∎

Example 4.58. The hand

$$Q\spadesuit\ 9\spadesuit\ Q\heartsuit\ 9\heartsuit\ A\diamondsuit\ 10\diamondsuit\ 8\diamondsuit$$

scores 29 points in diamonds, and should be backed with a Play bet. If the dealer holds

$$K\spadesuit\ K\heartsuit\ J\heartsuit\ Q\diamondsuit\ K\clubsuit\ 10\clubsuit\ 7\clubsuit,$$

her 27 in clubs loses to the player. The Blind bet pays off at 1–1. ∎

In a 52-card deck, the average card has a value of 7.38, so one might expect a typical 3 Card Blitz hand to total about 22. This does not account for the requirement that a valid hand be suited, of course, nor for the approximately 18.5% chance that the hand does not contain 3 suited cards (see page 153). With that in mind, strategy recommendations for 3 Card Blitz recommend playing on with a 20 or higher and folding with an 18 or lower [91].

That leaves 19s. Deciding whether or not to play with a 19 calls for counting how many points in other suits are in your hand and thus unavailable to the dealer—more is, of course, better for you. These are called *penalty points* [91]. If you hold 20 or more penalty points and a hand scoring 19, a raise is mathematically favored.

Example 4.59. If your hand is

$$2\spadesuit\ 6\heartsuit\ 4\diamondsuit\ 4\clubsuit\ Q\heartsuit\ 3\heartsuit\ 7\clubsuit,$$

then your best hand is $Q\spadesuit\ 6\spadesuit\ 3\spadesuit$, scoring 19. Adding up the rest of the hand gives 17 penalty points, and so the better choice is to fold this hand. Too many high cards remain in play.

If the $7\clubsuit$ is replaced by the $A\clubsuit$, then you have 21 penalty points, and should make the Play bet. ∎

To calculate the probability of drawing a 31 in at least one suit, we begin by computing

$$\frac{4\cdot\binom{4}{2}\cdot\binom{49}{4}}{\binom{52}{7}}\approx .0380.$$

This includes hands that might score 41 or 51, which need not be excluded, and the small number of hands that include two 31s. We have double-counted these hands, and so need to remove the probability of two 31s from this value. The probability of two 31s is

$$\frac{\binom{4}{2}\cdot\binom{4}{2}^2\cdot 46}{\binom{52}{7}}\approx 7.427\times 10^{-5}.$$

Subtracting gives

$$P(31)\approx .0379.$$

3 Card Blitz offers a simple Bonus bet, which pays off if the player's hand contains 4 or more cards of a single suit. The pay table for the Bonus bet is shown in Table 4.41.

TABLE 4.41: 3 Card Blitz: Bonus bet pay table

Flush length	Payoff
7 cards	200–1
6 cards	50–1
5 cards	8–1
4 cards	2–1

The probability of a 7-card hand with exactly k cards of a single suit, where $k\geqslant 4$, is

$$P(k)=\frac{4\cdot\binom{13}{k}\cdot\binom{39}{7-k}}{\binom{52}{7}}.$$

This function is tabulated below:

k	$P(k)$
4	.1954
5	.0285
6	.0020
7	5.131×10^{-5}

Summing these four values shows that the probability of winning a Bonus bet is .2259. Using this function, we can see that the Bonus bet carries a HA of 4.49%.

S7REAK

A product of the New Games Lab at the Crown Casino, *S7REAK* is a simple game based on the colors of up to 7 cards. Cards are dealt from a shoe so long as the color of the dealt cards is the same.

Example 4.60. If the first card is the 2♠ and the next is the 9♢, the hand stops there, with a 1-card Black streak.

If the first card is the 8♢, the cards are dealt until a black card is turned up. If the subsequent cards are 10♡ A♢ 5♢ 7♠, a Red streak of 4 cards has occurred. ∎

Players may bet on Red, Black, or No S7reak, and may make an optional Suit Up bet that takes suits into consideration. Red and Black pay off on the length of the streak, provided that the color matches. The payoffs are quite simple: a 1-card streak of the chosen color results in a push, and a streak of n cards pays off at $(n-1)$ to 1, up to the maximum 7-card streak, which pays 6–1.

No S7reak is just a simple even-money bet that the first 2 cards will be different colors. Assuming a 4-deck shoe for convenience, the probability of winning this bet as so written would be

$$\frac{208}{208} \cdot \frac{104}{207} > \frac{1}{2},$$

and the player would have the advantage. To return the edge to the casino, No S7reak pushes instead of winning if the first 2 cards form a pair, such as 7♢ 7♠. The probability of winning with a full shoe is then

$$\frac{104-8}{207} = \frac{96}{207} \approx .4638,$$

and the expectation of this bet is

$$E = (1) \cdot \frac{96}{207} + (-1) \cdot \frac{103}{207} = -\frac{7}{207} \approx .0338,$$

giving a HA of 3.38%.

Example 4.61. In the sequence 2♠ 9♢ dealt in Example 4.60, a Black bet would push, since the Black streak consisted of only 1 card. No S7reak would win, while Red would lose.

For the sequence 8♢ 10♡ A♢ 5♢ 7♠, Red would pay 3–1 on the 4-card streak, while Black and No S7reak would lose.   ∎

The Red and Black bets are essentially bets on the color of the first card, with an ascending payoff as further cards of the same color are dealt. As the length of the streak increases, of course, the probability that it will be extended decreases. Given a color, the probability of a streak in that color running for exactly k cards, $1 \leqslant k \leqslant 6$, is

$$P(k) = \frac{\binom{104}{k}}{\binom{208}{k}} \cdot \frac{104}{208 - k} = \left(\frac{104}{208} \cdot \frac{103}{207} \cdots \frac{105 - k}{209 - k} \right) \cdot \frac{104}{208 - k}.$$

In this formula, the factor $\dfrac{104}{208 - k}$ at the end denotes the probability of the last card being a different color, thus breaking the streak. For $k = 7$, this factor is not necessary since an eighth card is not dealt, and so

$$P(7) = \frac{\binom{104}{7}}{\binom{208}{7}} \approx .0070.$$

This function is shown in Table 4.42. We see that the probability drops by a

TABLE 4.42: S7REAK: Probability of a streak of k cards of one color

k	$P(k)$
1	.2512
2	.1256
3	.0625
4	.0309
5	.0152
6	.0075
7	.0070

bit more than ½ in moving from one card to the next, until the 7th and final card is dealt.

The simple payoff odds lead to a compact expression for the expectation of the Red and Black bets:

$$E = \sum_{k=1}^{7} [(k - 1) \cdot P(k)] - \frac{1}{2} \approx -.0160,$$

where the −½ term at the end represents the probability that the bet loses when the first card is of the opposite color. The HA is a very reasonable 1.6%.

4.4 Faro

In its late 19th-century heyday and into the early 20th century, *faro* was the dominant card game in gambling halls and saloons throughout western America. In faro's day, blackjack was a carnival game: the newcomer making inroads against the better-established game.

Game Play

Faro was a simple one-deck game where players bet on the rank of successively turned cards. Wagering was made easier by the sanctioned use of a *case keeper*: an abacus-like device on which a designated dealer kept track of the cards that had already been played. This constituted an early and open example of card counting. Figure 4.2 shows a case keeper. Beads are moved inward on the wires to show that a card of that rank has been dealt.

A turn of faro consisted of two cards dealt from the top of the deck, which was placed face-up in a wooden box. Bets were placed by putting chips on the corresponding card on the faro layout, which was a simple arrangement of one card of each rank. Frequently, the spade suit was depicted, as shown in Figure 4.3. The first card in the deck, called the *soda*, was exposed before any wagering took place and removed from play without any action. The next card was placed alongside the box, revealing the third card. A bet on the rank of the card placed to the side lost, while bets on the last exposed card won. The casino's advantage came on a *split*: when the two cards were of the same rank. In the case of a split, bets on that rank lost half the wager.

Example 4.62. Suppose that the cards so dealt were the 6♢ as the soda and the 3♠ to the side, exposing the 7♠. A bet on the rank 3 would lose, a bet on 7 would win, and bets on all other ranks, including the 6, would push. ∎

Players had the option of assigning their bet to the side card to win by capping their chips with a small hexagonal token or similar object. Often a copper penny was used, which led some players to call this "coppering" a bet. Figure 4.3 shows a faro layout with two wagers, one on the 4 to win and one on the 8 to lose.

All winning bets paid even money; once they were resolved, another round of betting followed and two more cards were exposed. This process ended after 24 turns, when the deck was down to 3 cards [27]. The last card in the deck was known as the *hock*; like the soda, it did not enter into rank wagering.

The final 3 cards were played out according to different rules. Since the case keeper showed the ranks of these cards, everyone at the table knew what remained in the box. The new bets available relied on bettors' ability to predict the order of the final 3 cards. If the 3 cards were all of different ranks, a gambler who correctly called their order, known as *calling the turn*, was paid

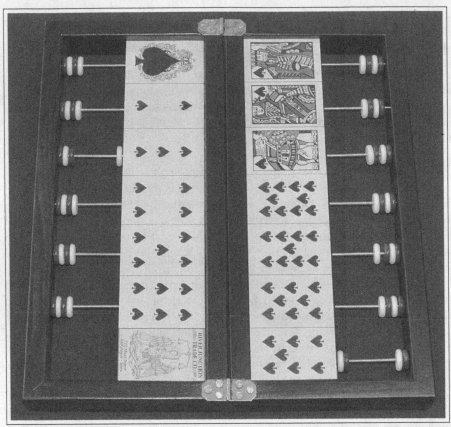

FIGURE 4.2: Faro case keeper. The beads show that two 7s and one 3 have been dealt out.

FIGURE 4.3: Faro layout. A chip is wagered on the 4 to win. The chip on the 8 has been topped with a small hexagonal token, indicating a bet on the 8 to lose.

at 4–1 [27]. Since there were $_3P_3 = 3! = 6$ different possible orders, the HA on this bet was 16.67%. If the final 3 cards included a pair, a configuration called a *cat-hop*, players were offered the chance to bet on where the odd card would fall among the 3. This even-money bet concealed a 33.33% house edge; if paid at 2–1, the bet was fair [27, 118].

Example 4.63. If the final 3 cards were all of the same rank, the hand ended early with no final round of betting. What is the probability of this event?

The chance that the last 3 cards are all of a specified rank is

$$\frac{\binom{4}{3}}{\binom{52}{3}} = \frac{4}{22,100} = \frac{1}{5225}.$$

Multiplying by 13 to cover all ranks gives a final probability of $\frac{1}{425}$—so this is unlikely. The faro proprietor had considerable opportunity to make money at the end of a deck on the high-edge wagers. ∎

Example 4.64. Consider a fresh deck where the soda's rank is known, and assume a wager on a different rank. The various probabilities applying to this bet are

$$P(\text{Lose}) = \frac{4}{51} \cdot \frac{47}{50} = \frac{94}{1275} \approx .0737$$

$$P(\text{Win}) = \frac{47}{51} \cdot \frac{4}{50} = \frac{94}{1275} \approx .0737$$

$$P(\text{Split}) = \frac{\binom{4}{2}}{\binom{51}{2}} = \frac{6}{1275} \approx .0047$$

$$P(\text{Push}) = \frac{1081}{1275} \approx .8478.$$

The winning and losing events cancel out in the expectation, making the expected value of this bet

$$E = \left(-\frac{1}{2}\right) \cdot \frac{6}{1275} = -\frac{6}{2550} \approx -.0024,$$

and the HA only .24%. ∎

Example 4.65. What is the edge on a first-draw wager made on the rank of the soda?

The probabilities noted above change as follows:

$$P(\text{Lose}) = \frac{3}{51} \cdot \frac{48}{50} = \frac{24}{425} \approx .0565$$

$$P(\text{Win}) = \frac{48}{51} \cdot \frac{3}{50} = \frac{24}{425} \approx .0565$$

$$P(\text{Split}) = \frac{\binom{3}{2}}{\binom{51}{2}} = \frac{1}{425} \approx .0024$$

$$P(\text{Push}) = \frac{376}{425} \approx .8847.$$

The house advantage can be calculated as above:

$$-\left[\left(-\frac{1}{2}\right) \cdot \frac{1}{425}\right] = \frac{1}{850} \approx .1176\%,$$

less than half the edge a bettor faces when betting on a different rank than the soda. ∎

The edges computed in Examples 4.64 and 4.65 presume that a wager that pushes is withdrawn after one turn, which the rules permitted. With the probability of a push running over 80%, it was not uncommon for faro players to let a wager ride through several 2-card turns until it was resolved. Since the probabilities change as cards are dealt, winning, losing, pushing, and splitting are not independent events from turn to turn. The case keeper made it easy to track a bet's chances as the cards were dealt.

Some faro houses included a "High Card" space at the top of the layout [54], as seen in Figure 4.3. The High Card bettor wagered that the winning card would be of a rank higher than the losing card, with the provision that aces were low. This bet could also be coppered to bet that the winning card would be the lower card. Splits on this bet were also handled by collecting half the wager. One possible appeal of this bet, for players and dealers alike, is that each round resolved the bet; there were no pushes.

Given the next two cards in the deck, each one had the same chance of being the winning card, so in the absence of a split, this was a fair bet. The exact HA of High Card depended on how many singleton cards remained in the deck; in the extreme case where only singletons were left and splits were impossible, the HA was 0%.

Changing Odds

As cards were dealt and the composition of the deck changed, the house's advantage changed. It was possible to lower the edge on some bets all the way to zero: when 3 cards of a given rank had been dealt, a bet on that rank, called a *case bet*, was then a fair wager—and the case keeper made it easy to tell when fair wagers were possible. Sharp faro players and casino managers knew this, of course, and so some gaming houses either required or expected players to make at least one negative expectation bet before placing fair wagers. The best (least negative) opportunity for players was a second-turn minimum bet on a rank with only 2 cards remaining; this arose when 2 of the first 3 cards matched in rank [107].

Example 4.66. What is the probability of a wager on the soda's rank lasting until the fourth turn and then resolving on that turn as a win or a loss (not a split)?

Each turn where the wager pushes sees the number of cards for the next turn drop by 2 while the number of soda-ranked cards remains 3. The probability of pushing on each of the first three turns is

$$\frac{\binom{48}{2}}{\binom{51}{2}} \cdot \frac{\binom{46}{2}}{\binom{49}{2}} \cdot \frac{\binom{44}{2}}{\binom{47}{2}} = \frac{2838}{4165} \approx .6814.$$

On turn 4, we are interested in the probability that exactly 1 of the 3 cards is drawn; this is

$$\frac{3 \cdot 42}{\binom{45}{2}} = \frac{7}{55} \approx .1273,$$

and so the final probability is

$$\frac{2838}{4165} \cdot \frac{7}{55} = \frac{258}{2975} \approx .0867,$$

less than 10%. ∎

In general, the probability of a soda bet running for n turns without resolution is

$$\prod_{k=0}^{2n-1} \left(\frac{48-k}{51-k}\right) = -\frac{(n-25)(2n-51)(2n-49)}{62,475}.$$

The probability of this bet then winning or losing on the next turn is

$$\frac{6(24-n)}{(25-n)(51-2n)},$$

so we have the product of these two probabilities, or

$$p_1(n) = \frac{6(24-n)(49-2n)}{62,475},$$

as the probability of a soda bet lasting exactly $n+1$ terms, $1 \leqslant n \leqslant 23$, and not splitting.

To compute the probability of a soda bet splitting on the $n+1^{\text{st}}$ turn, we simply replace

$$\frac{6(24-n)}{(25-n)(51-2n)},$$

with

$$\frac{3}{\binom{51-2n}{2}} = \frac{3}{(25-n)(51-2n)},$$

giving a split probability of

$$p_2(n) = \frac{3(49-2n)}{62,475}.$$

The functions $p_1(n)$ and $p_2(n)$ are tabulated in Table 4.43. We note that while these values are never very high, they reach their maxima at $n = 1$ and decrease thereafter.

TABLE 4.43: Faro probabilities for a soda bet to last $n + 1$ turns

n	$p_1(n)$	$p_2(n)$	n	$p_1(n)$	$p_2(n)$
1	.1038	.0023	13	.0243	.0011
2	.0951	.0022	14	.0202	.0010
3	.0867	.0021	15	.0164	.0009
4	.0788	.0020	16	.0131	.0008
5	.0712	.0019	17	.0101	.0007
6	.0640	.0018	18	.0075	.0006
7	.0571	.0017	19	.0053	.0005
8	.0507	.0016	20	.0035	.0004
9	.0447	.0015	21	.0020	.0003
10	.0390	.0014	22	.0010	.0002
11	.0337	.0013	23	.0003	.0001
12	.0288	.0012			

Returning to the soda bet described above, the probability that such a bet will be resolved on the fourth turn—win, lose, or split—is then

$$p_1(3) + p_2(3) \approx .0888.$$

Assessing the big picture asks that we look at the expectation of a bet that, once placed, is allowed to ride until it's resolved or the deck runs out, as it will if a case bet is placed and that card is revealed as the hock. For many faro players, the resolution of their bet, not the turn of 2 cards, was faro's fundamental unit [107]. From a betting perspective, a good faro strategy would include minimizing exposure to splits and exploiting case bets as much as possible. Mathematically, the 25 turns in a game of faro are expected to include, on average, 1.47 splits and the opportunity for 18.16 case bets, so this strategy had good potential to succeed [22].

Combination Bets

Some faro establishments offered players the chance to make other wagers, called *combination bets*, on more than one card rank at once. The available bets are shown in Figure 4.4.

FIGURE 4.4: Faro layout with multiple combination bets.

Combination bets are indicated by the alignment of the chip relative to the card image. The chips shown in Figure 4.4 are interpreted as follows:

- **A**: The simplest combination bet covered two adjacent cards and was made by placing a chip on the layout between the two cards. Chip A represents a bet on both the 5 and 6 to win. Some two-card combinations, such as the 8 and 2, were not available because the cards were not close enough together on the layout. This bet won if either of the two cards won, and lost if either card lost. Local custom dictated how this bet was handled when one card won and the other lost simultaneously: some establishments called this a push, while others declared a split and collected half the wager.

Other combination bets relied on careful chip placement [63].

- **B**: A chip placed on the inside corner (relative to the other cards) of a card is interpreted as a bet on that card as well as the two cards horizontally and vertically adjacent to it. In Figure 4.4, chip B, on the 5, also covers the 4 and 9.

- **C**: Placement of cornered chips is important. A chip placed touching an inside corner of a card, but not overlapping that card in any way, represents a bet on that card and the card diagonally adjacent to it in that direction. Chip C shows a combination bet on the 9 and 3.

- **D**: Chip D is placed on the outside corner of the ace—that is, away from the opposite row of cards. In this position, it represents a bet on the ace and the *second* card away in the direction of the corner where the chip is placed: the 3. One card is skipped, because there is another way to make a combination bet on two adjoining cards.

- **E**: Bets on 4 cards forming a 2 × 2 block are easily placed by putting a chip down in the region that they surround. Chip E is a bet on the ace, 2, queen, and king.

- **F**: The left side of the faro layout hosts a region called "the pot," where the 6, 7, and 8 come together. Chip F, placed in the pot, represents a combination bet on these three cards.

A pot wager carries different probabilities of resolution than a single-card bet; if the soda is the only exposed card and is known not to be a 6, 7, or 8, we have

$$P(\text{Lose}) = \frac{12}{51} \cdot \frac{39}{50} \approx .1835$$

$$P(\text{Win}) = \frac{39}{51} \cdot \frac{12}{50} \approx .1835$$

$$P(\text{Lose half}) = \frac{\binom{12}{2}}{\binom{51}{2}} \approx .0518$$

$$P(\text{Push}) \approx .5812.$$

Cheating at Faro

The low HA combined with lax gaming oversight in the 19th-century American West was a strong motivator for faro dealers to cheat. An obvious way of cheating was to stack the deck with multiple pairs of cards so as to activate the dealer's edge often. Some faro cheating was accomplished through the use of a crooked dealing box which allowed dealers to deal one or two cards as they wished. This was used in conjunction with a specially prepared deck of cards. Low cards were roughened up on the back with sandpaper, while high cards were roughened up on their faces. If a low card was followed by a high card, the two would tend to adhere on their rough surfaces, and so a skilled dealer could deal either one according to how the bets were running [87]. While not completely foolproof, it did allow a crooked game operator to affect the outcome. Other cheating schemes involved using the case keeper to record incorrect information regarding the cards that had been dealt.

In Germany, faro was dealt by nailing the shuffled deck to the table and tearing the cards away in turn; this certainly made cheating more difficult while driving up the dealer's expenses for fresh card decks [76].

Player cheating at faro was also possible, perhaps by quickly moving a bet from a neutral or losing card to the winning card after the cards were exposed, or by removing a copper when the first card of a pair was exposed. This was sometimes facilitated by attaching a fine horse hair to the token with wax, so that it could be quickly withdrawn with little fanfare and retracted into the crooked gambler's sleeve [76].

Faro retains the approval of the Nevada Gaming Control Board, and so may be legally offered in Nevada casinos [64]. While several Las Vegas casinos including the Aladdin, Dunes, Fremont, and Golden Nugget included faro on their gaming floors and had special chips manufactured just for the faro tables, the low HA relative to newer games eventually led to the game's disappearance from Nevada casinos. The last faro table in Las Vegas was at Palace Station and closed in the late 1970s; the last active game in Nevada was at the Ramada Casino in Reno and was removed in 1985 [108, 118].

Missed Opportunities?

Could more betting options, at a higher HA, have saved faro as a casino game? The main bets at craps, Pass and Don't Pass, have HAs less than 1.5%, lower if free odds are added to the wager. A standard craps table also offers a host of other bets at a range of higher HAs, as high as 16.67%, and newer craps betting options such as the Fire Bet take the casino's edge even higher. Doing the same for faro seems like an achievable goal.

One possibility might be a Split bet, which would pay off if the next two cards form a pair—any pair. With only the soda removed, the probability that the second and third cards in a fresh deck have the same rank is

$$\frac{12 \cdot \binom{4}{2} + \binom{3}{2}}{\binom{51}{2}} = \frac{75}{1275} = \frac{1}{17},$$

so payoff odds of 15–1, which might attract enough action to make this a viable option, would give the casino a 5.88% advantage.

This edge would change with the composition of the deck, and the use of a case keeper would ensure that gamblers would have access to that information. In an extreme case where only 1 or 0 cards were left in all ranks, the HA would be 100%; a casino would do well by its patrons to take no action on splits when that happened.

Another option could be a suit-based bet. Since suits are not tracked on the case keeper, the casino would not be providing information of use to players. One possibility would be to offer a bet on the suit of the hock. This bet faces 3–1 odds against; if paid at 2–1, the HA would be 25%—perhaps too high. However, if the bet could be placed late in the game, players who devoted minimal effort to tracking suits might be able to turn this bet in their favor. If the last three cards were all known to be of the same suit, players' could guarantee a win, so some sort of limit on how late this bet could be made would be necessary.

Consider a hypothetical bet on the suit of the last ace to be dealt. (We choose the ace for convenience; this wager could be introduced on any card rank.) As with the suit of the hock bet above, the odds against this bet are 3–1 when placed at the start of the deck. Paying a winning bet at 5–2 makes the HA 12.5%. While this bet could also be introduced on the suit of the first card of any rank to appear, assigning the wager to the last ace ensures that players remain at the table, where they might possibly make other bets, until their bet is resolved.

A number of bets from the Pell layout (page 143) might be plausible additional bets for faro, but those bets were easily attacked by card counters playing Pell with two 6-deck shoes. The risk from counters, especially in a game with faro's tradition of keeping the count for the players, would be much greater in single-deck faro.

Pharaon

Faro's history dates back to a game called *pharaon*, which was popular in 17th-century France and takes its name from the pharaoh depicted as one of the kings on some early playing cards. Pharaon differed from faro in the following ways [22]:

- Once placed, a bet had to remain on the layout until it was resolved.

- The deck was held and dealt face down, so there was no soda or hock, and a complete game lasted 26 turns rather than 25. Calling the turn was not an option, since the last turn contained 2 cards, not 3.

- The last 2 cards were handled differently: A split on the final turn meant that the house took the player's entire bet instead of half, and on a case bet, if the player's card was the last in the deck, this was a push, not a win.

These differences in the rules—turning half a house win into a win and a house loss into a push—worked to increase the house advantage slightly. This advantage was quantified by French mathematician Abraham de Moivre in 1756 and examined in detail by Swiss mathematician Leonhard Euler in 1764 [15, 23]. Their work showed, as one might reasonably expect, that the exact HA depended on the number x of "significant" cards, or cards of the rank upon which the player was betting that were still in the deck, and the total number of cards n remaining to be dealt.

- If $x = 1$, a recursive set of equations, similar to the analysis used to compute the Top Rung HA on page 150, showed that the house advantage was simply $\frac{1}{n}$. This quantity ranges from $\frac{1}{48} \approx 2.08\%$ to $\frac{1}{2} = 50\%$, since n must be even and at least 3 cards must have been dealt to reduce x to 1.

This result may also be found by noting that the HA with 1 significant card is derived from the rule that a match on the last card is a push, not a win. Otherwise, the significant card is just as likely to generate a player win as a house win. Since the probability that the significant card is last in the deck is $\frac{1}{n}$, this is also the house edge.

- If $x = 2$, the HA was found by both men to be $\frac{n+2}{2n(n-1)}$. This is less than $\frac{1}{n}$ whenever $n > 4$, so the house has a higher edge if there is 1 significant card remaining rather than 2 unless there are very few cards left to be dealt.

- Similar formulas can be derived for $x = 3$ and $x = 4$ [15, 23]. Moreover, though pharaon did not permit combination bets, Euler continued his analysis up through $x = 8$ as an academic exercise. The results are shown in Table 4.44.

TABLE 4.44: Pharaon house advantages with n cards to be dealt [23]

x	HA
1	$\dfrac{1}{n}$
2	$\dfrac{n+2}{2n(n-1)}$
3	$\dfrac{3}{4(n-1)}$
4	$\dfrac{2n-5}{2(n-1)(n-3)}$
5	$\dfrac{5n-10}{4(n-1)(n-3)}$
6	$\dfrac{3(2n^2-13n+16)}{4(n-1)(n-3)(n-5)}$
7	$\dfrac{7(2n^2-12n+13)}{8(n-1)(n-3)(n-5)}$
8	$\dfrac{4n^3-50n^2+176n-151}{2(n-1)(n-3)(n-5)(n-7)}$

Additionally, Euler noted that, if the cases $x = 1$ and $x = 2$ were disregarded as irregularities, a general formula for the HA as a function of n and

x was

$$\frac{x}{2^x} \left(\frac{x-1}{1(n-1)} + \frac{(x-1)(x-2)(x-3)}{1\cdot2\cdot3\cdot(n-3)} + \frac{(x-1)(x-2)\cdots(x-5)}{1\cdot2\cdot3\cdot4\cdot5\cdot(n-5)} + \cdots \right).$$

This could be used to derive corresponding formulas for the more complicated combination bets that would later be introduced to faro.

Example 4.67. For the pot wager covering all 6s, 7s, and 8s, $x = 12$ and we have a HA of

$$\frac{12}{4096} \left(\frac{11}{n-1} + \frac{165}{n-3} + \frac{462}{n-5} + \frac{330}{n-7} + \frac{55}{n-9} + \frac{1}{n-11} \right)$$

$$= \frac{3(4n^5 - 122n^4 + 1360n^3 - 6715n^2 + 13,906n - 8598)}{4(n-11)(n-9)(n-7)(n-5)(n-3)(n-1)}.$$

As with the standard wagers considered by Euler, this HA is at a minimum when n is at its highest and a maximum when n is lowest, ranging from 6.46% when $n = 52$ up to 50% when only 12 cards—all of them necessarily 6s, 7s, or 8s—remain to be dealt. In this case, the next hand is guaranteed to be a split, and the house will take half the wager.   ■

4.5 Red Dog

Red Dog, which is also known as Acey-Deucy or Yablon, might be described as an "anti-matching" game: the object for the bettor, with one uncommon exception, is to avoid matching cards [69]. Players make an initial bet, and then two cards, from a single deck or multi-deck shoe, are dealt to the table.

- If the cards are of the same rank, a third card is immediately dealt. If this card forms three of a kind, all bets are paid off at 11–1. If the third card is a different rank than the first two, the bets push.

- If the cards have consecutive ranks, as with a 6 and a 7, then all bets push and the hand ends without dealing a third card.

- Otherwise, the "spread" of the two cards, or the number of ranks between them, is determined and announced by the dealer. With aces counting 14, kings 13, queens 12, jacks 11, and all other cards their face value, the spread s of two cards with ranks x and y is $s = |x - y| - 1$. (Identical-rank cards have a spread of -1 using this formula, but there is no need to calculate s in that circumstance.) Note that aces always rank high in Red Dog.

TABLE 4.45: Red Dog payoff odds [69]

Spread	Payoff odds
1	5–1
2	4–1
3	2–1
4–11	1–1

Players now have the option of doubling their bets. A third card is dealt. If it falls in between the first two cards, players are paid at fixed odds which depend on the spread, as shown in Table 4.45. If the third card does not fall in between the first two, including the case where the third card pairs either of the first two, all player bets lose.

The option to raise one's bet after the spread is known introduces player choice to Red Dog, which suggests that an optimal strategy for doubling exists. Since twice as much money is being risked, doubling is the better move if the spread offers at least a 50% chance of winning.

Let $s > 0$ denote the spread (so the first two cards are assumed to be neither the same rank nor of consecutive ranks) and n the number of decks in use. If the hand is dealt from a fresh n-deck shoe, the probability of winning a hand of Red Dog is

$$p = \frac{4ns}{52n - 2} = s \cdot \frac{4n}{52n - 2}.$$

The fraction involving n is approximately $\frac{1}{13}$, with the approximation improving as n increases, so a good estimate of p is

$$p \approx \frac{s}{13},$$

from which it follows that if the spread is at least 7, $p > .5$ and doubling the bet is the correct decision.

If $s = 7$, the probability of winning ranges from .5600 in a single-deck game to .5411 if 8 decks are used. The highest probability of winning if $s = 6$ is .4800, in a single-deck game.

How likely is this event? Assuming that all two-card pairs are equally likely, there are $13 \cdot 13 = 169$ ways to draw two cards, looking only at the ranks. 30 of these pairs lead to a spread of 7 or higher, making the probability of a spread that justifies doubling

$$\frac{30}{169} \approx .1775,$$

which means that this opportunity arises on slightly less than 20% of all hands.

Example 4.68. Doubling a bet only if $s \geqslant 7$ means accepting only even money payoff odds on doubled bets. If the spread is 1 and the payoff is 5–1, what is the house edge as a function of n if you double your bet?

The expectation of a doubled bet is

$$E = (10) \cdot \frac{4n}{52n - 2} + (-2) \cdot \frac{48n - 2}{52n - 2} = \frac{4 - 56n}{52n - 2}.$$

In a single-deck game, this is –$1.04, so the HA on a $2 wager is 52%. The HA is also 52% if you do not double your bet, so it's better to keep the bet at its initial level and limit your losses.

As n increases, this expectation approaches $-\dfrac{14}{13} \approx -\$1.0769$, and the HA approaches 53.85%. This strongly supports the intuitive choice not to double a bet if $s = 1$. ∎

Successive hands of Red Dog are not independent, and so the game might seem to be susceptible to card counting. The cards that a counter should track are the extreme ones: 2s, aces, 3s, and kings. These cards, while they might give rise to large spreads which can be profitable, also are unlikely to fall in between the first two dealt cards: a 2 or ace on the third card is a guaranteed loser for players. Removing all of the 2s and aces from an n-deck shoe changes the probability of winning with a spread of s to

$$\frac{4ns}{44n - 2},$$

which is greater than ½ when $s \geqslant 6$, and so there are slightly more opportunities to double a bet in a favorable deck configuration. If all of the aces and 2s are gone, the probability that $s \geqslant 6$, favoring a double, is .2479.

Does card counting matter? In the scenario above with all 48 of the aces and 2s removed from a 6-deck shoe, the player has an edge of .175%, but standard playing conditions make it unlikely that this advantage can be achieved [90]. If 75% of the shoe (234 cards) is customarily dealt out, the probability of exhausting the aces and 2s before reshuffling is

$$\frac{\binom{264}{186}}{\binom{312}{234}} \approx 2.5578 \times 10^{-7}.$$

One could expect to play through 3,909,636 shoes before encountering one with even this meager advantage.

At the other extreme, the worst case for the gambler would be a shoe with all of the 8s removed, as 8 lies in the exact center of the Red Dog deck. As above, the probability of dealing out all of the 8s in 234 cards is

$$\frac{\binom{288}{210}}{\binom{312}{234}} \approx 7.3345 \times 10^{-4}.$$

While a player could play through an average of 1363 shoes before such a deal occurs, this is small compensation for a HA of 3.49% [90].

4.6   Casino Cribbage

Cribbage is a two- to four-player card game, most commonly played by 2 players, that dates back to the 17th century [27]. Points are tracked on a small board with pegs. In 2014, Casino Cribbage, stripped of some head-to-head game elements and the board, was launched as a new carnival game.

In standard two-hand cribbage, each player receives 6 cards and must discard 2 to a side hand, the "crib," which belongs to the dealer. Each player benefits from an upcard turned over at the start of play and common to all hands. The game is played in two rounds: one round where players alternate playing cards from their 4-card hands trying to reach a combined total of 15 or 31, with additional points scored for playing consecutive cards of the same rank or "runs" of cards in sequence, and a second round where players count the number of points in their hands, with points scored for combinations totaling 15 and for sets of cards of the same rank, in sequence, or suited.

Cribbage is won by the first player to score 121 points over several hands. Casino Cribbage is a simplified one-hand version of the game where players seek to outscore the dealer's hand in a simple point count. Players and dealer are each dealt 5 cards and must discard one to make a 4-card hand. A player cuts the remaining cards, and a community card is turned up, which is held in common by all player and dealer hands. The dealer has an edge in that he can see the community card before discarding, and so can maximize his hand's value with full knowledge of the cards in play. Players, in discarding, must make their best choice in the face of incomplete information about what the community card might be.

In the showdown, players win if their hands contain more points than the dealer's. Hands are valued by counting various combinations of cards:

- Any combination of 2 or more cards adding up to 15 scores 2 points. Face cards count 10, all other cards count their value.

- Any run of 3 or more cards in sequence scores the number of consecutive cards.

- A pair scores 2 points. 3 of a kind (a *pair royal*) scores 6, since 3 pairs may be formed. 4 of a kind (*double pair royal*) scores 12, 2 points for each of the $\binom{4}{2} = 6$ pairs.

- Flushes of 4 or 5 cards score the number of cards in the flush.

- One point is scored if a hand holds the jack of the suit of the community card. This is called *nobs*.

Example 4.69. Suppose the player holds

$$J\heartsuit \ 2\heartsuit \ 3\diamondsuit \ J\diamondsuit$$

and that the community card is the 7$\diamondsuit$. This hand scores 2 points for $J\heartsuit$ 2$\heartsuit$ 3$\diamondsuit$, 2 for $J\diamondsuit$ 2$\heartsuit$ 3$\diamondsuit$, 2 for the pair of jacks, and 1 for nobs, totaling 7 points.   ∎

Example 4.70. The hand

$$7\diamondsuit \; 7\clubsuit \; 8\spadesuit \; 8\diamondsuit$$

together with a 6$\heartsuit$ as community card scores 24 points.

- There are 4 combinations of an 8 and a 7 adding up to 15, totaling 8 points.

- Four runs of 3 consisting of the 6, one of the 7s, and one of the 8s total 12 points.

- The pair of 7s counts 2 points, as does the pair of 8s.

                                                                    ∎

The highest possible hand of 29 is achieved when holding three 5s and a jack, with the community card the 5 of the jack's suit. This hand scores 8 points from the four 15s formed by 3 of the 5s, 8 points from the four 15s formed by combining a 5 with the jack, 12 points from the four 5s, and 1 point for nobs.

As an illustration of the players' and dealer's thought processes, and the dealer's advantage, suppose that the player has been dealt

$$7\spadesuit \; 3\heartsuit \; 6\clubsuit \; 9\spadesuit \; 7\diamondsuit.$$

There are 4 points in this hand: 2 for the pair of 7s and 2 for the 15 formed from the 9 and 6.

- If the 3 is discarded, the hand improves if the community card is an ace, 2, 5, 6, 7, 8, or 9; a total of 24 cards of the 47 remaining. The probability of improvement is then $\frac{24}{47} \approx .5106$.

- Removing the 6 leaves a hand that improves when an ace, 3, 6, 7, 8, or 9 is drawn. 19 cards improve this hand, but at a cost of 2 points that were lost when the 6 was discarded. Drawing an ace, 6, or 9 merely restores the 2 points that were sacrificed with the 6.

- Discarding a 7 also forfeits 2 points, and leaves a hand that improves on a 2, 3, 5, 6, 7, 8 or 9: only 22 cards improve this hand, and only 13 improve it by more than the 2 lost points.

- Finally, dropping the 9 leaves a hand improved by an ace, 2, 3, 5, 6, 7, 8, or 9. There are 27 cards that improve the hand, but again, this sacrifices 2 certain points, and drawing an ace, 3, or 9 as the community card merely restores the hand to its value before discarding.

The best move for the player is probably to throw the 3; this leaves a good chance of improvement without immediately hurting the low-scoring hand. Since the goal of Casino Cribbage is to beat the dealer's hand, throwing away points without a substantial chance of gaining back more than was lost with the community card is risky behavior.

If the community card is the 10$\diamondsuit$ and the dealer holds

$$9\clubsuit \ 10\spadesuit \ Q\clubsuit \ K\clubsuit \ 4\heartsuit,$$

the best she can do is throw away any card other than the 10$\spadesuit$, since there's no way to get more than 2 points from those 6 cards. We see that the dealer's full knowledge does not translate into victory; while the player gains nothing from the 10$\diamondsuit$, he nonetheless wins, 4–2.

If the player's hand outscores the dealer's, payoff odds are based on the value of the player's hand, in accordance with Table 4.46. Hands with the same score push. Since it is impossible to deal a 5-card hand totaling 19, 25, 26, or 27, these totals are omitted from the table. "I have 19" is a classic way for a cribbage player to announce that his or her hand scores 0.

TABLE 4.46: Casino Cribbage pay table [11]

Player's hand	Payoff odds
1–12	1–1
13–16	2–1
17, 18, 20	3–1
21–24	10–1
28	100–1
29	1000–1

An optional Casino Cribbage bet was the "Nobs or Heels" bet, which paid off at 5–1 if a jack was turned up as the community card (in standard cribbage, this is referred to as "heels," and the dealer scores 2 points) or if the player making the bet held the nobs. A simple approach to the community card probability says "There are 52 cards in a standard deck, and 4 of them are jacks, so the probability of the community card being a jack is just 4 out of 52, or 1 out of 13."

A more complicated look at that question might say "But that probability's not constant. We need to account for how many jacks are held by the player and the dealer."

It turns out that the more complicated look leads to the same simple answer: $\frac{1}{13}$. For simplicity in this example, assume one player going head-to-head against the dealer. The conditional probability that the community card is a jack, given that the player's and dealer's hands together contain k jacks, where $0 \leqslant k \leqslant 4$, is

$$P(\text{Jack} \mid k \text{ jacks}) = \frac{4-k}{42}.$$

We need not be concerned with how the jacks are distributed between the two hands since we're just looking for the number that have been dealt among the 10 cards, not where they landed.

The probability that the 10 cards held by the player and dealer contain exactly k jacks between them is

$$P(k) = \frac{\binom{4}{k} \cdot \binom{48}{10-k}}{\binom{52}{10}}.$$

From the formula for conditional probability, we have

$$P(k \text{ jacks dealt and a community jack}) = P(\text{Jack} \,|\, k \text{ jacks}) \cdot P(k)$$

$$= \frac{4-k}{42} \cdot \frac{\binom{4}{k} \cdot \binom{48}{10-k}}{\binom{52}{10}}.$$

This factor must be summed from $k = 0$ to $k = 4$, which yields

$$P(\text{Jack as the community card}) = \sum_{k=0}^{4} \frac{4-k}{42} \cdot \frac{\binom{4}{k} \cdot \binom{48}{10-k}}{\binom{52}{10}} = \frac{1}{13},$$

confirming the simple analysis with which we began. Essentially, what happens here is that a low probability of dealing k jacks is balanced by a decreased probability of turning over a jack as the community card once k jacks are in the hands. Moreover, the probability that no jacks are dealt, giving the highest chance of drawing a community jack, is over 41%, which represents a significant contribution to the sum.

If instead there are n hands including the dealer's, we have

$$P(\text{Jack as the community card}) = \sum_{k=0}^{4} \frac{4-k}{52-5n} \cdot \frac{\binom{4}{k} \cdot \binom{48}{5n-k}}{\binom{52}{5n}},$$

which also gives $\dfrac{1}{13}$.

A major limiting factor faced by Casino Cribbage was that potential players needed to be familiar with the standard rules of cribbage, which sharply restricted the potential audience.

4.7   Card Games with Nonstandard Decks

Casino game designers are generally advised not to tamper too much with standard casino equipment. Where card games are concerned, this advice can be reduced to "Aside from maybe adding a joker, or possibly using a Spanish deck with the 10s removed, don't change the standard 52-card deck." The designers of the games in this section went against that advice.

Casino Dominoes

In the 2007 movie *Ocean's 13*, Bernie Mac's character plays a dealer running a game called 'Nuff Said, which is a version of dominoes adapted for the casino floor. While the movie does not describe 'Nuff Said in any detail, *Casino Dominoes* brought dominoes to the Plaza Casino in 2018 as a card game. This game was created by Harold Moret in conjunction with the Center for Gaming Innovation at the University of Nevada, Las Vegas.

A full set of standard "double six" dominoes includes 28 tiles: each individual domino bears 0 through 6 spots on each of its 2 sides. Order does not matter: ⚃ and ⚃ are the same domino. The 28 dominoes include 7 "doubles" with the same number on each side and $\binom{7}{2} = 21$ whose sides have different numbers of spots. For ease in dealing and gameplay, Casino Dominoes replaced the actual dominoes by playing cards with domino images printed on them. The game is dealt from a combined deck containing two complete sets, or 56 cards.

The classic game of dominoes calls for players to match domino numbers and extend a chain; a doubles domino may be placed at right angles to the chain and allows for the possibility of two new chains. When the exposed ends sum to a multiple of 5, the player receives that number of points. Figure 4.5 shows a chain with actual dominoes.

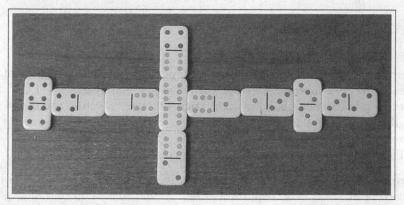

FIGURE 4.5: In-progress game of dominoes. The game began with the 6–6 domino.

Players make a main bet and an optional Doubles bet, and are then dealt 3 cards and one designated "connector" card. The challenge is to match any or all of the cards in the 3-card hand to the connector, individually, in such a way that the unmatched ends add up to a multiple of 5. A non-doubles domino can be matched with the connector leaving 2 free ends; a doubles tile is matched at a right angle to the connector and has 3 scoring ends. Unlike in dominoes, no skill is involved in play.

Example 4.71. Suppose that a player is dealt the ⚀ , ⚅ , and ⚁ with the connector the   .

Only the cards bearing numbers which also appear on the connector have the potential to score points, so the   cannot score. The   can be joined to the connector at their   ends, but the sum of the exposed numbers is 7, which is not divisible by 5, so no points are scored. Placing the   card at right angles to the   end of the connector gives 3 ends adding up to 5 points. This hand scores a total of 5.   ∎

The main bet is paid off in accordance with Table 4.47.

TABLE 4.47: Casino Dominoes pay table

Points	Payoff odds
40	500–1
35	200–1
30	50–1
25	25–1
20	12–1
15	8–1
10	3–1
5	2–1

Example 4.72. The maximum score of 40 points is achieved when the player is dealt a   connector together with two   s and a   . Each   scores 15 points when placed at right angles to the   end of the connector, and the   scores 10 points when its   end is joined to the   on the connector.

There are 4 ways to draw this winning combination.   ∎

Example 4.73. A ⚃ connector makes it impossible to score any points: connecting at the   end requires another   end to add up to a multiple of 5, and the   card must be placed at right angles, leaving 3 scoring ends totaling 7. Similar problems occur when trying to join a card at the   end.

If the player's connector card is the   , then he or she receives an immediate 2–1 payoff on the main bet. The probability of this payoff is

$$\frac{2}{56} \approx .0357,$$

which adds $7\frac{1}{7}$¢ to the expectation of the main wager.   ∎

This payoff is not awarded for a   dealt to the player's 3-card hand, although   can score no points from the hand either, no matter what the connector card is.

The optional Doubles bet pays on the number of doubles dominoes in the player's hand. Table 4.48 shows the payoff odds. The "Double Doubles"

TABLE 4.48: Casino Dominoes: Doubles bet pay table

Player's hand	Payoff odds
Maiden's Hand	300–1
Double Doubles	100–1
4 Doubles	20–1
3 Doubles	4–1
2 Doubles	2–1

payoff is won when the player has two pairs of two different doubles, for example ⚋ , , ⚋ , and . This particular hand scores 0 points in play regardless of which card is the connector; a 100–1 payoff if the Doubles bet was made is a pretty good consolation prize. The probability of landing Double Doubles from a fresh 56-card deck is

$$\frac{\binom{7}{2}}{\binom{56}{4}} = \frac{21}{367,290} = \frac{1}{17,490}.$$

The Maiden's Hand is a 4-card hand containing the 4 highest doubles: ⚏ , ⚏ , ⚏ , and . The hand takes its name from a legend of a suitor who won a maiden's hand in marriage by drawing 4 winning tiles from a set. This hand, which also scores 0 points in play, has probability

$$\frac{2^4}{\binom{56}{4}} = \frac{16}{367,290} \approx \frac{1}{22,956}.$$

A 50-Card Poker Deck

In 2013, Empire Global Gaming, Inc. began marketing a 50-card deck designed to provide new opportunities in old card games [20]. This new deck consisted of aces through tens in five suits or colors.

We consider the implications of this new deck on five-card poker. With no face cards, royal flushes are impossible, and they are replaced as the best possible hand by five of a kind. There are ten of these in a 50-card deck, making the probability of five of a kind

$$P(\text{Five of a kind}) = \frac{10}{\binom{50}{5}} = \frac{10}{2,118,760} = \frac{1}{211,876}.$$

Five-of-a-kinds in a 50-card deck are thus about three times more likely than royal flushes in a 52-card deck. Of course, neither hand has all that high a probability.

Straight flushes can be counted just as they are in a standard deck, by focusing on the lowest-ranked card in the straight flush. Since aces are both high and low in a 50-card deck, any card from ace through seven can anchor a straight flush; there are 35 of these. Once the lowest card is chosen, the other 4 cards are determined, and so the probability of a straight flush is

$$P(\text{Straight flush}) = \frac{35}{2,118,760} = \frac{1}{60,536}.$$

Continuing down through the list of standard poker hands reveals several differences from poker with a standard deck. For example, flushes are far less common with the 50-card deck than the standard deck, and four-of-a-kinds are more common. A moment's thought makes this reasonable: in moving from 4 suits of 13 cards each to 5 suits of 10 cards, we are diminishing the number of cards in any one suit by 3 and increasing by 1 the number of cards in the deck of a given rank.

Flushes are counted very simply: There are 5 suits to choose from, and once a suit is selected, 10 cards from which 5 are chosen. From this collection of one-suit hands, we need to subtract the 35 straight flushes found above, and so the number of flushes in a 50-card deck is

$$5 \cdot \binom{10}{5} - 35 = 1225.$$

For four-of-a-kinds, we begin by choosing the rank from the 10 available, and then choose 4 of the 5 cards of that rank. The fifth, nonmatching card may then be any of the 45 remaining cards in the deck, giving a total of

$$10 \cdot \binom{5}{4} \cdot 45 = 2250$$

four-of-a-kinds.

The frequencies of the remaining 5-card poker hands in a 50-card-deck are compiled in Table 4.49.

That flushes outrank straights in the 50-card deck is no surprise, as we have seen that flushes outrank straights in a standard deck. Two surprises emerge from close inspection of this table:

- Flushes are over 7 times rarer than full houses. In a standard deck, full houses are less common than flushes and so outrank them.

- The most likely hand rank is one pair; high-card-only hands are less common than one-pair hands.

A new possibility brought about by the fifth suit is a hand where all five suits are represented—this is called a *rainbow* hand. Quick counting will show that there are $10^5 = 100,000$ possible rainbow hands. Five of a kind is trivially a rainbow hand, while flushes and straight flushes cannot be rainbow hands.

TABLE 4.49: Poker hand frequencies: 50-card deck

Hand	Frequency
Five of a kind	10
Straight flush	35
Flush	1225
Four of a kind	2250
Full house	9000
Straight	21,840
Three of a kind	90,000
Two pairs	180,000
High card	764,000
One pair	1,050,000

Example 4.74. Rainbow straights, may represent a new type of hand worth enumerating. For convenience in this example, let the fifth suit be stars, denoted by ★.

As with straight flushes, we can count rainbow straights by starting with the lowest-valued card, and again, this is any card from ace through 7: 35 in all. For example, suppose that this card is the $A\clubsuit$. There are then four choices for the second card in a rainbow straight: the $2\diamondsuit, 2\heartsuit, 2\spadesuit$, and $2★$. Once this card is chosen, another suit has been eliminated and only three choices remain for the third card. Following this pattern to the last card, we see that the total number of rainbow straights is then

$$35 \cdot 4 \cdot 3 \cdot 2 \cdot 1 = 840.$$

∎

If rainbow straights are separated out as a new kind of hand, reference to Table 4.49 shows that a rainbow straight would properly beat a flush but lose to a straight flush, and that removing these 840 hands from the roster of straights would not change the ranking of a straight relative to the other hands. The new hand frequency distribution is given in Table 4.50.

Picture Perfect

Picture Perfect uses a 52-card deck with the aces, 2s, and 3s removed and replaced by extra jacks, queens, and kings. Players and dealer are each dealt six cards, and a player wins if their hand contains more face cards than the dealer's, similar to High-Card Flush.

There are 24 face cards in a Picture Perfect deck; the number of ways to draw a 6-card hand including k face cards is the product of two terms: the number of ways to draw k face cards from 24 and the number of ways to draw

TABLE 4.50: Poker hand frequencies: 50-card deck with rainbow straights considered separately

Hand	Frequency
Five of a kind	10
Straight flush	35
Rainbow straight	840
Flush	1225
Four of a kind	2250
Full house	9000
Straight	21,000
Three of a kind	90,000
Two pairs	180,000
High card	764,000
One pair	1,050,000

$6 - k$ non-face cards from the remaining 28. The probability of a hand with exactly k face cards is

$$p(k) = \frac{\binom{24}{k} \cdot \binom{28}{6-k}}{\binom{52}{6}};$$

this is similar to the formula for counting matches at Rouleno, where we must account for both "hits" and "misses." This gives rise to the probability distribution in Table 4.51.

TABLE 4.51: Probability distribution for Picture Perfect

k	$p(k)$
0	.0185
1	.1159
2	.2776
3	.3257
4	.1973
5	.0585
6	.0067

Evaluation of the sum

$$\sum_{k=0}^{6} k \cdot p(k)$$

shows that the mean number of face cards in a Picture Perfect hand is about 2.77.

Since players are not permitted to discard and replace any of their cards, the strategy element of Picture Perfect comes from the player's choices among the four available bets:

- The **Ante** bet must be made on every hand before the cards are dealt.

- The **Faces** bet, equal to the Ante, is also required on every hand before the deal. To win this bet, a player must beat the dealer and have at least 4 face cards in his or her hand. Table 4.52 shows the pay table for Faces.

TABLE 4.52: Pay table for the Picture Perfect Faces bet [38]. Payoffs require that the player's hand beat the dealer's

Face cards	Payoff
6	25–1
5	3–1
4	1–1
3	Push

- The optional **Picture Perfect** bet pays off if a hand contains three face cards of the same rank, or two face cards of the same rank and suit. This bet only pays off on hands that beat the dealer, though. Table 4.53 shows the payoffs.

TABLE 4.53: Pay table for the optional Picture Perfect bet [38]

Cards in one rank	Payoff
6	1000–1
5	200–1
4	20–1
3	3–1
2, if suited	2–1

- Once the cards are dealt, players may examine their hands and either make a **Play** bet equal to the Ante bet, or fold, forfeiting both the Ante and Faces bets.

If the dealer qualifies with at least two face cards ranking king-high or better, including a pair of jacks or queens, then the hands are compared. If the player's hand beats the dealer's, the Ante bet is paid off at even money and the Play bet pays 2–1. If the dealer fails to qualify, the Ante bet pushes but the Play bet remains active.

Example 4.75. Suppose that you are dealt

$$6\heartsuit \; J\heartsuit \; 8\clubsuit \; 4\heartsuit \; Q\spadesuit \; K\heartsuit.$$

You hold 3 face cards, but do not win on the Picture Perfect bet, if you made it. You choose to make the Play bet. If the dealer's cards are

$$9\heartsuit \; 10\clubsuit \; Q\spadesuit \; 4\diamondsuit \; Q\diamondsuit \; Q\heartsuit,$$

then she qualifies and also holds 3 face cards. Hands with equal numbers of face cards such as these are compared by looking at the ranks of the cards, so your KQJ beats the dealer's QQQ. You win even money on your Ante, 2–1 on your Play bet, and push the Faces bet. ∎

The probability that the dealer qualifies is

$$p(6) + p(5) + p(4) + p(3) + p(2) - P(QJ).$$

The last term is the probability that the dealer holds a queen, a jack, and no other face cards, and so fails to qualify despite holding two face cards.

Since there are 8 queens, 8 jacks, and 28 non-face cards in a Picture Perfect deck, we have

$$P(QJ) = \frac{8 \cdot 8 \cdot \binom{28}{4}}{\binom{52}{6}} \approx .0644.$$

Using this value together with the values from Table 4.51 gives a probability of .8013 that the dealer qualifies. An 80% qualification rate seems reasonable; it can be adjusted by changing the qualifying criteria.

- If the dealer qualifies on any 2 face cards, the probability of qualifying rises to 86.6%.

- If the dealer requires 3 face cards to qualify, she qualifies only 58.8% of the time. This is too low to retain players. After losing out on some Ante bet payoffs on strong hands where the dealer failed to qualify, players would seek out better games.

Super Easy Aces

In a hypothetical casino with a robust selection of carnival games, the surplus aces, 2s, and 3s removed from Picture Perfect decks could have been put to use at an adjacent table where *Super Easy Aces* was played. This game uses a 54-card deck with 25 aces, 12 deuces, 8 threes, 7 fours, and 2 jokers. Suits have no standing in Super Easy Aces except for one bet that pays off an extra unit if a $2\diamondsuit$ or $3\diamondsuit$ is dealt; for this reason, 2s and 3s are evenly distributed among the 4 suits.

Super Easy Aces was described by its inventor, Paul Harry, in 2018 as a combination of roulette, Casino War, and craps [9].

- As in roulette, players try to predict what the result of the next turn will be. In roulette, it's the spin of a ball on a wheel; in Super Easy

Aces, it's the deal of a single card. This is the game's main bet, and it pays off as shown in Table 4.54.

TABLE 4.54: Super Easy Aces pay table

Rank	Payoff	Probability
Ace	1–1	25/54
Deuce	3–1	9/54
2♦	4–1	3/54
3	5–1	6/54
3♦	6–1	2/54
Joker	25–1	2/54

The payoffs on a dealt 2♦ or 3♦ represent an even-money bonus paid to winning 2 or 3 bettors; these are not separate wagers. All ace, deuce, and 3 bets push if the next card dealt is a joker.

- As in Casino War, players and dealer are dealt one card each. An optional Dealer Match side bet pays off varying amounts, from 1–1 to 100–1, if the player's card matches the dealer's card.

- As in craps, the 4 is a loser for all main bets—it is not possible to bet that your next card will be a 4. The 4 functions like a 7 in craps after the shooter has established a point. A player covering all available ranks on the Super Easy Aces layout is not guaranteed a win, much like a craps bettor with Place bets on every point loses them all when the shooter rolls a 7.

The Dealer Match bet does pay off when both player and dealer are dealt 4s; this is the only case where a player's dealt 4 leads to a win. The pay table for this bet is shown in Table 4.55.

TABLE 4.55: Super Easy Aces: Dealer Match pay table

Rank	Payoff
Ace	1–1
Deuce	3–1
3	5–1
4	8–1
Joker	50–1

Example 4.76. Of the four primary bets—ace, deuce, 3, and joker—which gives the lowest HA?

The expected values of each $1 bet are these:

Ace: $E = (1) \cdot \dfrac{25}{54} + (-1) \cdot \dfrac{27}{54} = -\dfrac{2}{54} \approx -\$.0370.$

Deuce: $E = (3) \cdot \dfrac{9}{54} + (4) \cdot \dfrac{3}{54} + (-1) \cdot \dfrac{40}{54} = -\dfrac{1}{54} \approx -\$.0185.$

3: $E = (5) \cdot \dfrac{6}{54} + (6) \cdot \dfrac{2}{54} + (-1) \cdot \dfrac{44}{54} = -\dfrac{2}{54} \approx -\$.0370.$

Joker: $E = (25) \cdot \dfrac{2}{54} + (-1) \cdot \dfrac{52}{54} = -\dfrac{2}{54} \approx -\$.0370.$

The best bet for the player is the Deuce bet, whose HA is half that of the other three bets. ∎

Absent the 1-unit Diamond Bonus on the Deuce and 3 bets, their HAs would both rise, from 1.85% and 3.70%, respectively, to 7.40%. This bonus is desirable to bring the four bets' HAs into closer alignment with one another and balance the game.

Example 4.77. What would be the new HA on the Ace bet if it also carried a 1–1 bonus, on the $A\diamondsuit$?

Since there are 25 aces in a Super Easy Aces deck, it is not possible to distribute them evenly across all four suits. If 6 aces are diamonds, we have

$$E = (1) \cdot \frac{19}{54} + (2) \cdot \frac{6}{54} + (-1) \cdot \frac{27}{54} = \frac{4}{54} \approx \$.0740,$$

a 7.40% player advantage that is untenable, so we need to dismiss any notion of a suit-balanced subset of aces within the deck.

Let x denote the number of $A\diamondsuit$s among the 25 aces in the deck. Such a deck gives an expected value of

$$E = (1) \cdot \frac{25-x}{54} + (2) \cdot \frac{x}{54} + (-1) \cdot \frac{27}{54} = \frac{x-2}{54}$$

for the Ace bet; this value is only negative if $x = 1$, which yields the same HA as the Deuce bet: 1.85%. ∎

Since including only one $A\diamondsuit$ among 25 aces would likely be noticed by players and cause them eventually to avoid the Ace bet due to the infrequent bonus, there is no bonus on the Ace bet.

Dragon Poker

With a possible nod toward the millennial market (gamblers born roughly between 1980 and 2000) and collectible card games such as Magic: The Gathering, *Dragon Poker* entered a field trial at the Rio Casino in 2018. While the

Dragon Poker deck had 53 cards including one wild card, the Gold Dragon, this was far from a "standard-52-card-deck-plus-a-joker." The cards of the Dragon Poker deck are listed in Table 4.56. Notice that the cards, except for

TABLE 4.56: Dragon Poker's 53-card deck

Card	Count
Gold Dragon (wild)	1
Fire Dragon	3
Water Dragon	3
Phoenix	7
Tiger	8
Panda	9
Monkey	10
Rabbit	12

the dragons, have only one identifier instead of separate ranks and suits. For most gameplay purposes, the Fire and Water dragons are interchangeable; the different types of dragons only matter when the optional Dragon Bonus bet is considered.

The basic Dragon Poker game is a variation on 3-Card Poker. With the absence of both suits and numerically ordered ranks, the only possible hands are three of a kind, one pair, and three unmatched cards. Hands are ranked within these three classes by the relative scarcity of the cards, so 3 tigers beats 3 pandas but loses to 3 phoenixes.

Example 4.78. The probability of a particular triple, given that n is the number of cards in that rank, is

$$\frac{\binom{n}{3} + \binom{n}{2}}{\binom{53}{3}},$$

where the two terms in the numerator count "natural" triples and jokered triples, respectively. The fundamental combinatorial identity

$$\binom{n}{2} + \binom{n}{3} = \binom{n+1}{3}$$

allows us to simplify this formula to

$$\frac{\binom{n+1}{3}}{\binom{53}{3}},$$

which is easily interpreted as the chance of drawing 3 cards from a pool of $n + 1$, including the gold dragon, from a 53-card deck. Using this formula for dragons requires that we treat the gold dragon as a separate kind of card and take $n = 6$.

The probability of any three of a kind is then found by summing the 6 values given by this expression, which gives approximately .0318.   ∎

Play begins with equal Ante and Ante Bonus bets, after which the players and dealer are each dealt 3 cards face down. After inspecting their cards, each player may choose to fold or to match their Ante bet with a Raise bet. Players staying in by raising then face off against the dealer, with the higher hand winning. The dealer has no minimum qualifying hand; all hands see action. Winning Ante and Raise bets pay 1–1. The Ante Bonus bet pays according to Table 4.57 on the strength of the player's hand, but *only* if that hand beats the dealer. If the player and dealer tie, all bets push.

TABLE 4.57: Dragon Poker: Ante Bonus pay table

Hand	Payoff
3 Dragons	20–1
3 Phoenixes	7–1
Any other triple	4–1
Pair, Pandas or higher	1–1
Winning hand less than 2 Pandas	Push

Example 4.79. A player dealt two pandas and a dragon would do well to raise. If the dealer then revealed a tiger, a monkey, and a rabbit, the player would win even money on the Ante and Raise bets and 2–1 on the Ante Bonus.   ∎

Since there is no dealer qualifying hand, the house edge lies in the Ante Bonus bet. Because that bet loses on a strong hand that is beaten by the dealer's stronger hand, some hands of a pair of pandas or better cannot cash in on the payoff.

Example 4.80. A $1 player holding 3 rabbits who is beaten by 3 monkeys loses $3. If the Ante Bonus bet was not in force, she would lose $1 on the Ante and Raise bets while winning $4 on the Ante Bonus for a net gain of $2.   ∎

Two optional bets are available: the 3 Card Bonus and the Dragon Bonus. The first of these pays off on the strength of a player's hand without the requirement that that hand beat the dealer: this is a kind of "bad beat" insurance. If a player's hand of 2 tigers or higher loses to an even better dealer hand, the Ante Bonus may be lost, but the 3 Card Bonus still pays off. Table 4.58 shows the payoff odds.

The Dragon Bonus bet is the only place where the different types of dragon are important. This simple wager pays off on the number and kind of dragons in a player's hand. Payoffs start at 5 for 1 for a hand containing the Gold Dragon alone, with the top payoff of 1000 for 1 going to a hand with 3 Fire Dragons or 3 Water Dragons without the Gold Dragon—see Table 4.59.

TABLE 4.58: Dragon Poker: 3 Card Bonus bet pay table

Hand	Payoff
3 Dragons	40–1
3 Phoenixes	30–1
3 Tigers	15–1
Any other triple	10–1
2 Dragons	4–1
2 Phoenixes	2–1
2 Tigers	1–1

TABLE 4.59: Dragon Poker: Dragon bonus bet pay table

Hand	Payoff
3 Fire Dragons	1000 for 1
3 Water Dragons	1000 for 1
3 Dragons with Gold Dragon	200 for 1
Any 3 Dragons	60 for 1
Any 2 Dragons	7 for 1
Gold Dragon	5 for 1

4.8 Exercises

Solutions begin on page 288.

Poker

1. In Four-Card Poker, count the number of possible four-card straight flushes and four-of-a-kinds when six cards are dealt, and so confirm that four of a kind beats a straight flush for the dealer as well.

2. In Two-Card Joker Poker, how many possible 10-high hands are there?

3. An alternate pay table for Dueling for Dollars' Two-Card Poker Bonus bet is given in Table 4.60.

Find the HA for the Two-Card Poker Bonus bet with 5, 6, and 8 decks using this pay table.

4. A second alternate Two-Card Poker Bonus pay table is given in Table 4.61. Here, the bet loses if an unsuited pair is drawn, but the payoffs on a straight or straight flush are increased. A suited pair, such as 4♣ 4♣, is considered as a flush.

TABLE 4.60: Dueling for Dollars: Two-Card Poker Bonus pay table #2 [113]

Hand	Payoff odds
Straight flush	2.5–1
Straight	1–1
Flush	1–1
Pair	2–1

TABLE 4.61: Dueling for Dollars: Two-Card Poker Bonus pay table #3 [113]

Hand	Payoff odds
Straight flush	5–1
Straight	2–1
Flush	1–1

a. Find the new probabilities of drawing a flush in 2 cards as a function of the number of decks n.

b. Find the HA for this bet with 5, 6, and 8 decks.

5. When playing Mississippi Stud, find the probability that you are dealt a guaranteed winning hand (not merely a nonlosing hand) in your first two cards.

6. In a game of Wild Hold 'Em Fold 'Em, you are dealt 4♣ 4♡ 7♡.

a. Should you back this hand up with a Bet wager, or fold?

b. If you make the Bet wager and are dealt the $A\diamondsuit$, should you go on to make the Raise wager, or fold?

7. A Lunar Poker player holding two pairs is advised simply to play on without buying a sixth card. There are two ways that a sixth card can generate a second winning hand: by turning 2 pairs into a full house or by pairing the odd card to create a second two-pair hand. Find the probability that a sixth card improves a 2-pair hand.

8. Lunar Poker's Instant Payoff provides for a variety of payoff options. The maximum payoff odds are 500–1 on a straight flush and 1500–1 on a royal flush [100]. How much of an advantage does this give a player?

9. In example 4.26 (page 127), the HA of the Poker-All High and Low bets is found to be 15.41%. How does this compare to the HA on the similar-looking Any 5 Red and Any 5 Black bets, which pay 8–1 if 5 or 6 of the balls are of the color selected?

10. For the Poker-All suit bet described on page 128, find the probability p of winning.

11. The original patent application for Poker-All described a game where the casino drew 5 balls, rather than the 6 that were drawn at the Sahara Reno [30]. This, of course, changes the probabilities, and the payoff odds changed to reflect this. Find the expected value and HA of the following Poker-All bets with the casino drawing 5 balls.

a. A Flush bet paying 200 for 1 if all 5 balls were the same suit. It is not necessary for the bettor to choose a suit.

b. 5 Low cards, which paid 32 for 1.

c. "All Red," which paid 30 for 1 if all 5 balls were red cards.

d. "All Face Cards," a bet not offered at the Sahara that paid 400 for 1 on a hand with 5 face-card balls.

12. If Poker-All is played by drawing 5 balls instead of 6, does the 8–1 bet on the 8◊ still carry a player advantage?

13. Find the probability of the following Pokerbo hands, assuming that the hand is dealt from the top of a 6-deck shoe.

a. Four of a kind.

b. Full house.

c. Two pairs.

14. Using the result of Exercise 13, find the house advantage on the Pokerbo Two Pairs bet.

15. Suppose that two standard decks were used in a 5-card poker game. This opens up the possibility of some new hands, while changing the probabilities of standard hands.

a. Which hand has a lower probability (and thus ranks higher): a royal flush or five of a kind?

b. Find the probability of a flush with 2 pairs.

c. Suppose that flushes with 2 pairs are separated out as a different type of hand, so a "flush" in double-deck poker includes only flushes with one or no pairs. Under this interpretation, does a full house beat a flush?

16. *Badugi* is a variation on draw poker where the object is to hold the lowest 4-card poker hand. Poker games with this goal are referred to as *lowball* poker. Players are initially dealt 4 cards and have three opportunities, separated by rounds of betting, to discard and replace any or all of their cards in a quest for low-ranking hands [34].

As is customary in lowball poker, straights and flushes have no standing, but in badugi, hands with all four suits represented rank lower, and thus better, than hands with 2 or more cards in a single suit. For example, the hand 7$\diamondsuit$ 6$\heartsuit$ 4$\spadesuit$ 2$\clubsuit$ beats 6$\clubsuit$ 5$\clubsuit$ 3$\spadesuit$ A$\diamondsuit$, even though the second hand ranks lower as a high-card hand without looking at suits. A four-suited hand is called a *badugi*. The lowest hand is then a four-suited 432A. Competing badugis are ranked by comparing their highest cards, with the lower highest card winning, as when a 9732 badugi outranks a Q543 badugi.

a. If your initial hand is K$\heartsuit$ K$\diamondsuit$ 7$\heartsuit$ A$\diamondsuit$, find the probability of drawing into a badugi in one turn by discarding both kings.

b. Starting from a fresh 52-card deck, find the probability of being dealt a badugi.

Combination and Matching Games

17. In a game of Triple Shot, confirm the probabilities shown in Table 4.35 (page 137) for the following poker hands, with 6 cards dealt to each hand.

a. Full house

b. Flush

c. Three of a kind

d. Pair of queens, kings, or aces

18. a. Find the probability that the first 20 cards dealt in a round of Show Pai contain no 10s or face cards.

b. If 20 Show Pai cards have been dealt with no 10-count cards showing, the probability of the player receiving four 10s on her next hand has fallen from 1 in 117 to 1 in x. Find x, to the nearest integer.

19. Confirm the assertion on page 143 that the Black-Black and Red-Red Pell bets are fair bets if the "house wins all bets on two identical cards" rule is not included. Assume two full six-deck shoes.

20. Ubetcha Games has published a house edge for the Color bet at Top Rung of about 5.5% [110]. Use the infinite deck approximation to find the approximate HA. How does it compare to Ubetcha's value?

21. What 3 Card Blitz hand leads to the lowest score? Find the probability of receiving that hand.

22. An optional Blitz Bonus bet may be added to 3 Card Blitz, as the casino may choose. The SLS Casino in Las Vegas (now the Sahara) offered this choice in 2019. This bet pays off on the player's hand, according to Table 4.62.

a. Find the probability of a 7-card hand containing a 5-card royal flush.

b. Find the probability of a Double Blitz.

TABLE 4.62: Pay table for 3 Card Blitz's Blitz Bonus bet

Hand	Payoff odds
5-card royal flush	2500–1
Double Blitz	250–1
Royal Blitz	25–1
Blitz	10–1
Score of 30	5–1

Faro and Red Dog

23. a. Find the probability that 25 cards are dealt from a standard deck with no 6s, 7s, or 8s appearing.

b. In faro, compute the probabilities of the four possible outcomes of a pot wager made after 12 turns and the soda are complete with no 6s, 7s, or 8s dealt.

24. The rules of faro permit players to bet on some cards to win and some to lose on the same turn. A player who simultaneously loses on both cards on one turn is said to be *whipsawed*. If the soda is the $10\heartsuit$ and a gambler has placed a pot wager to win and a combination bet on the king and queen to lose, find the probability that she is whipsawed.

25. In faro, repeatedly betting on a single card rank which loses 4 times in one deal of 25 hands is called *losing out* [2]. Find the probability of losing out. (Betting on a single card rank which wins 4 times in one deal is called *winning out*, and has the same probability.)

26. *Short Faro* is a variation on faro that uses only the 9s through aces in 2 or 3 standard decks [2]. Three cards are initially dealt face down: these are the "house" cards. Players bet on any of the 6 card ranks, and 3 more cards are dealt face up. A bet wins if at least one card of that rank is exposed; payoffs are 1–1 if only 1 card shows the chosen rank, 2–1 if two show it, and 3–1 if all three do. The payoff structure is the same as that seen in the 3-die game chuck-a-luck (page 229), but the ranks of the three player cards are not independent, which changes the expectation.

a. Construct the probability distribution function for a hand of Short Faro using 3 decks (72 cards).

b. What is the mean number of cards a gambler can expect to match?

c. Find the expected value of a $1 bet when Short Faro is dealt from 3 decks.

d. If a gambler is paid 4–1 instead of 3–1 when all three cards are of the chosen rank, find the new HA.

27. For n running from 1 through 8, find the probability of an 11–1 payoff in Red Dog when the first three cards dealt from an n-deck shoe have the same rank.

Games with Nonstandard Decks

Casino Dominoes

28. Find two different ways to draw a 35-point Casino Dominoes hand.

29. Find the maximum possible score in a Casino Dominoes hand with a connector.

30. Calculate the house advantage of the Casino Dominoes Doubles bet.

31. We noted in Example 4.71 that only the cards bearing numbers which also appear on the connector have the potential to score points.

a. Assuming that your connector shows two different numbers, find the probability that none of your 3 cards contains either of those numbers.

b. Assuming that your connector is a doubles domino, find the probability that none of your 3 cards also shows that number.

A 50-Card Deck

32. In a game of five-card poker with a 50-card deck, find the number of rainbow poker hands with the following ranks:

a. Four of a kind

b. Full house

c. Two pairs

Picture Perfect

33. In a hand of Picture Perfect, suppose that you are dealt 6 jacks. Find the probability that you lose to the dealer's hand.

Tapis Vert

Tapis Vert (French: "green carpet") was a card-based matching game operated by the French national lottery from 1987–1993. The game used a 32-card deck, with 7s through aces in all 4 suits [57]. Players selected one card from each suit, and the lottery drew one card from each suit in a televised daily drawing. Tickets matching 2 or more cards were paid according to Table 4.63.

TABLE 4.63: Pay table for Tapis Vert [57]

Cards matched	Payoff odds
2	1–1
3	30–1
4	1000–1

34. a. In the drawing of 29 March 1988, the lottery drew all four aces [3]. This was consistently the most popular combination chosen by lottery players, and the cumulative payoff to winners greatly exceeded the money taken in for the drawing. Following that drawing, lottery officials moved to cap the amount of action permitted on a single set of 4 cards. Find the probability of matching all 4 drawn cards.

b. Show that the lottery held a 40% advantage on a Tapis Vert ticket.

c. While 40% is a routine house edge for lottery games, it is far too high to be viable in a casino carnival game. Devise a new pay table for Tapis Vert that pays off on 2–4 matches but has a house advantage between 2% and 6%.

d. Repeat part c. for a 5-suit deck paying off on 2–5 matched cards.

Chapter 5

Dice Games

Standard casino dice are precision-machined: perfect cubes to within .0001 inches (.00254 millimeters) on each side. The spots on the dice are made of a different-colored plastic with the same density as that used for the dice and are inlaid flush with the surface rather than being indented, as is the case with common game dice. Moreover, casino dice have razor-sharp edges and corners, in order to ensure that they bounce and roll in a truly random fashion across the 14 feet of a standard casino craps table. Common board game dice usually have rounded edges and corners; this means that they can roll more and generate reasonably random rolls in a space the size of a game board.

Some carnival games have experimented with dice having more than six sides. These games include Spider Craps (page 29) and Super Color Sic Bo (page 238). The ancient Greeks knew that there were only 5 regular polyhedra: solid figures whose faces are a single type of regular polygon, with the same number of faces meeting at each vertex. These solids are known as the *Platonic solids*, and have 4, 6 (a standard cube), 8, 12, or 20 sides. Figure 5.1 shows each of these 5 platonic dice.

FIGURE 5.1: Platonic dice with 4, 6, 8, 12, and 20 sides.

It is customary to denote a die with x sides as a "dx," as in a d6 or d8 for a 6- or 8-sided die. Two six-sided dice are similarly denoted by writing 2d6.

5.1   1-Die Games

Die Rich

Die Rich overcame what many regarded as an unfortunate name [45, 50] to find a place on the floor at the Luxor Casino in Las Vegas in 2006. Die Rich is played with one standard six-sided die, and is designed to be completed in no more than four rolls. This efficiency is a selling point; in traditional craps, it theoretically can, and frequently does, take a large number of rolls to complete a round and resolve a table full of wagers.

Die Rich is played as follows:

- On the first roll, 6 wins and 1 loses. If a 6 is rolled, the payoff is 1–1.

- If neither a 1 nor a 6 is rolled, the number rolled becomes the *point*, as in craps, and the shooter rolls up to three more times.

    - If a 1 comes up on any of these three rolls, the shooter loses immediately.

    - If the point comes up on the first or last of these rolls, the payoff is 2–1.

    - If the point comes up on the second roll, the payoff is 1–1.

The tree diagram in Figure 5.2 illustrates the various paths that can be taken from the first roll to the resolution of a $1 bet. A round of Die Rich begins at the left side of this figure.

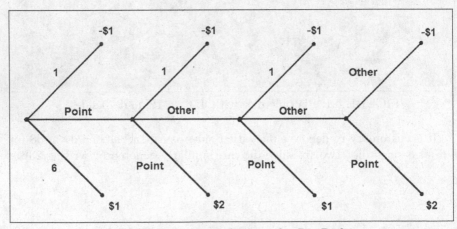

FIGURE 5.2: Tree diagram for Die Rich.

The probability of winning a Die Rich bet is the sum of four probabilities, each corresponding to one of the downward-pointing branches on the tree in

Figure 5.2. We define a *neutral* result on the second through fourth rolls as a result that is neither a 1 nor the established point. The four possible winning probabilities are these:

- $p_1 = P(6 \text{ on the first roll})$
- $p_2 = P(\text{Point on the first roll, point on the second})$
- $p_3 = P(\text{Point on the first roll, neutral second roll, point on the third})$
- $p_4 = P(\text{Point on the first roll, two neutral rolls, point on the fourth})$

Since successive Die Rich rolls are independent, each of these probabilities can be computed by finding the probability of each edge leading to the associated outcome and multiplying those values together. The probability of rolling a point on the first roll is $\frac{4}{6}$, since four of the six sides of the die bear a possible point number. Once a point is set, the probability of re-rolling it drops, to $\frac{1}{6}$. We have

$$p_1 = \frac{1}{6}$$
$$p_2 = \frac{4}{6} \cdot \frac{1}{6} = \frac{1}{9}$$
$$p_3 = \frac{4}{6} \cdot \frac{4}{6} \cdot \frac{1}{6} = \frac{2}{27}$$
$$p_4 = \frac{4}{6} \cdot \left(\frac{4}{6}\right)^2 \cdot \frac{1}{6} = \frac{4}{81}.$$

Adding shows that

$$P(\text{Win}) = p_1 + p_2 + p_3 + p_4 = \frac{65}{162} \approx .4012,$$

and the Complement Rule gives

$$P(\text{Lose}) = 1 - P(\text{Win}) = \frac{97}{162} \approx .5988.$$

The expected value of a \$1 Die Rich bet is

$$E = p_1 + 2p_2 + p_3 + 2p_4 + (-1) \cdot \frac{97}{162} \approx -\$.0370,$$

giving a low HA of 3.70%. While this may have made the game appealing to some gamblers, it is nonetheless greater than the HA for craps bets such as Pass and Don't Pass.

As an exercise in game design, we consider the effect on the Die Rich HA of changing the payoff values.

- Paying off all wins at even money raises the house edge to 19.75%, greater than even the worst standard bet on a craps table: Any Seven, which carries a 16.67% casino advantage.

- Raising the payoff for re-rolling the shooter's point to 2–1 on all of the last three rolls switches the 3.70% edge from the casino to the gambler, and thus renders the game undesirable to casino table games managers.

- One viable alternate Die Rich pay table offers 2–1 for a win with a 6 on the first roll, and even money for setting and making a point thereafter. The HA for this version is 3.09%.

- Another pay table that reduces the HA but still favors the house is 1–1 for a win on the first or second roll, 2–1 for a win on the third, and 3–1 for a win on the fourth and final roll. This pay structure, which rewards success more if it comes on the later rolls, leads to a HA of 2.47%.

Die Rich offered two optional bets separate from the main game. Players could at any time make a one-roll bet on any number from 1–6, and were paid 4–1 if their number was thrown on the next roll. This bet was mathematically equivalent to Any Seven, carrying a 16.67% HA. The Shooter's Streak bet allowed players to profit from a shooter who successively made 2 or more points in a row. The pay table for Shooter's Streak is shown in Table 5.1.

TABLE 5.1: Pay table for the Shooter's Streak bet at Die Rich

Points Made	Payoff odds
2	1–1
3	3–1
4	6–1
5	10–1
6	30–1
7	60–1
8	100–1
9	300–1
10	600–1
11 or more	1000–1

The probability of a gambler making exactly n points before failing to make the next point, for $0 \leqslant n \leqslant 10$, is simply

$$P(n) = p^n \cdot (1 - p),$$

where $p = \dfrac{65}{162}$ is the probability of winning computed above. Since the payoffs are capped after 11 wins, the probability of winning the top prize is $P(11) = p^{11}$. This actually computes the probability that the shooter will make 11 or more consecutive points.

Let It Ride

As a card game, Let It Ride has had a respectable run among casinos' carnival game lineups [6]. As a 1-die game, *Let It Ride* (LIR) is a thought experiment developed by gaming writer David Sklansky in an attempt to illustrate unexplored possibilities for new games [102].

Let It Ride was conceived as a game that would allow players the chance at a large payoff for a small wager, as is available when buying a lottery ticket or in a casino when playing keno, but with more excitement than lotteries and keno typically afford. Rather than waiting for 20 numbers to be drawn in the range 1–80, which lacks action compared to blackjack or craps, Let It Ride gave gamblers a more exciting game offering a chance at a big jackpot. The game involved playing a small initial bet over multiple successive wins of a game without requiring great risk of a player's accumulated winnings.

Figure 5.3 shows the proposed Let It Ride layout. Each space on the layout .

123	124	125	126	134
135	136	145	146	156
234	235	236	245	246
256	345	346	356	456

FIGURE 5.3: Let It Ride layout.

contains 3 of the numbers from 1–6; there are $\binom{6}{3} = 20$ wagering options. A player chooses 3 numbers on a die and places a chip in the corresponding space. This is a wager that one of those 3 numbers will show on successive tosses of a single die. Assuming a fair die, each of these spaces carries the same probability of success, ½, on every roll.

The minimum LIR wager was $2; the player's initial bet was all the money he or she risked in a single game played to its resolution. If any one of the player's 3 numbers was rolled, they won the roll. As a string of winning rolls unfolded, the player's chip would be capped with a "progress token" noting the number of consecutive wins in the current streak [102]. Once 3 straight winning rolls were recorded, the gambler was insured against loss: successive rolls comprised a sequence of even-money wagers where the player was guaranteed their previous level of winnings when the streak of wins ended.

Example 5.1. Consider a $2 wager on 125 through the following sequence of rolls.

First roll: 2.

Second roll: 1.

Third roll: 2. At this point, the player will not lose, even if the next roll is a 3, 4, or 6.

Fourth roll: 5. The player will win at least $2.

Fifth roll: 5. The 8 for 2 payoff guarantees a profit of at least $6.

Sixth roll: 5. The streak continues. The new guaranteed payoff is $16.

Seventh roll: 2.

Eighth roll: 2.

Ninth roll: 2.

Tenth roll: 1.

Eleventh roll: 2.

Twelfth roll: 4. The streak ends, and the player receives $512, a $510 profit. ■

Payoffs were based on the number of successive winning rolls, beginning with an even-money payoff for 3 winning rolls and doubling with each additional win until reaching the maximum payoff of $8192 on 15 straight wins. Algebraically, the payoff for k consecutive winning throws, $k \geqslant 3$, was simply 2^{k-2} for 2.

Example 5.2. A player who had won on 7 straight throws won $32 for their $2 wager. Though his or her action continued to the next roll, if that roll lost, the payoff was still $32—the win was guaranteed and could only grow, without further funds at risk. ■

The probability of winning exactly k straight rolls is

$$P(k) = \left(\frac{1}{2}\right)^k \cdot \frac{1}{2} = \frac{1}{2^{k+1}},$$

since we must include the last factor, the probability of losing on the $k + 1^{\text{st}}$ roll.

The probability of winning something on a round of LIR was then

$$\sum_{k=3}^{\infty} \left(\frac{1}{2}\right)^{k+1} = 1 - [P(0) + P(1) + P(2)] = 1 - \left[\frac{1}{2} + \frac{1}{4} + \frac{1}{8}\right] = \frac{1}{8}.$$

Since Let It Ride payoffs successively doubled while the probability of winning at each level dropped by half each time, computing the expectation is simple. LIR payoffs stop at 15 straight wins, so the probability of winning $8192 does not include the final extra factor of ½. The payoffs and probabilities are listed in Table 5.2.

TABLE 5.2: Probability distribution and pay table for Let It Ride

Number of wins	Probability	Payoff (for 2)
0	.5000	0
1	.2500	0
2	.1250	0
3	.0625	2
4	.0313	4
5	.0156	8
6	.0078	16
7	.0039	32
8	.0020	64
9	9.76×10^{-4}	128
10	4.88×10^{-4}	256
11	2.44×10^{-4}	512
12	1.22×10^{-4}	1024
13	6.10×10^{-5}	2048
14	3.05×10^{-5}	4096
15	3.05×10^{-5}	8192
$k, 2 < k < 15$	$\dfrac{1}{2^{k+1}}$	2^{k-2}

The expected value of a \$2 minimum wager is

$$E = \sum_{k=3}^{14} \left[(2^{k-2}) \cdot \frac{1}{2^{k+1}} \right] + (8192) \cdot \frac{1}{2^{15}} - 2$$

$$= \sum_{k=3}^{14} \left[\frac{1}{8} \right] + \frac{2}{8} - \frac{16}{8}$$

$$= -\$0.25.$$

The HA is found by dividing by \$2: 12.50%.

The cap on winnings of \$8192 was necessary to preserve a casino advantage. Since the probabilities and payoffs combine to give a constant product of $\frac{1}{8}$, extending Table 5.2 by two more lines, to a maximum payoff of \$32,768 for 17 straight wins, would produce a fair game, with a HA of 0. Another step taken in the design of LIR that insulated the casino against the possibility of huge losses was a maximum bet of \$8 per square [102]. With this rule in force, the maximum casino loss on any one streak of rolls would be \$65,536. This assumes that all 3 numbers assigned to a specific square are rolled at some point in the streak, so that only 1 of the 20 squares qualifies for the maximum payoff. In an extreme counterexample, a run of 15 straight 3s could potentially cost the casino $65,536 for each of the 10 squares where a 3 appears, or $655,360 in total.

No Lose Free Roll

While Let It Ride never made it to casinos, the idea of a wager that could ride without risk once it had won a specified number of times resurfaced in the No Lose Free Roll craps side bet offered at Vegas World in the 1980s. The connection is Sklansky, who served as a gaming consultant—his official job title was "Resident Wizard"—to Vegas World owner Bob Stupak, and may have had a hand in designing that bet [101].

No Lose Free Roll was a variation on the craps Field bet. "No Lose" was, of course, a misnomer, as the bet had a $\frac{26}{36}$ probability of losing. For the lucky gambler who bucked these over 2-to-1 odds against winning, the potential for great riches loomed—or so the promotional materials indicated.

A standard Field bet covers the numbers 2, 3, 4, 9, 10, 11, and 12, with bonus payments of 2–1 on 2 or 12, and occasionally a 3–1 payoff on 12. The No Lose Free Roll took some combinations away from the bettor and added in some others: the bet paid off at even money if the following roll was a 2, 3, 11, or 12, as well as a hard-way 4, 6, 8, or 10 [71].

Example 5.3. The house advantage on a $5 bet on the standard Field bet is 5.56%. What is the house edge *on the first roll* of a $5 No Lose Free Roll wager?

The first roll either wins at even money, with probability $p = \frac{10}{36}$, or loses, with probability $q = 1 - p = \frac{26}{36}$. The expectation is then

$$E = (5) \cdot \frac{10}{36} + (-5) \cdot \frac{26}{36} = -\frac{80}{36} \approx -\$2.22,$$

which, when divided by the $5 initial bet, gives a very high house advantage of 44.44%, eight times higher than the HA on the original Field bet. ∎

Ten of the 36 possible rolls were winners, while 26 were losers. How, then, could this be billed as a "no lose" bet? The attraction, as with Let It Ride, was the ability, once the No Lose Free Roll bet had won once, to let one's winnings ride for subsequent rolls, without risk of loss. Winnings could accumulate until the first losing roll, at which time all of a player's collected chips were returned. There was a ceiling on accumulated winnings: an initial bet of $5 had a maximum payout of $20,000, for example. This means that the $5 win in Example 5.3 was a minimum, as the player had the potential to win far more if the designated numbers kept coming up on the dice.

To reach this maximum, a $5 wager would have to hit 12 times in a row, for a total win of $20,480—only $20,000 of which would be paid out. The probability of this event is

$$\left(\frac{10}{36}\right)^{12} = \frac{1,000,000,000,000}{4,738,381,338,321,616,896} \approx 2.1104 \times 10^{-7},$$

approximately once in every 4,738,381 games.

This is, as one might expect, unlikely, but what if you could string together a shorter run of lucky rolls? How does the No Lose Free Roll, in its entirety, compare to other craps side bets, whose house edges run as high as 16.67%?

The amounts to be won on this roll are \$5, \$15, \$35, ... \$10,235, \$20,000, or $10 \cdot 2^{n-1} - 5$ as n runs from 1 to 11, followed by a final payoff of \$20,000. To win $10 \cdot 2^{n-1} - 5$, it is necessary to hit a winning number on n consecutive rolls and follow that up by rolling a losing number, which gives

$$P(10 \cdot 2^{n-1} - 5) = p^n \cdot q.$$

With $p = \dfrac{10}{36}$ and $q = \dfrac{26}{36}$, the expectation is then

$$E = \sum_{n=1}^{11} \left[(10 \cdot 2^{n-1} - 5) \cdot \left(\frac{10}{36}\right)^n \cdot \left(\frac{26}{36}\right) \right] + (20,000) \cdot \left(\frac{10}{36}\right)^{12} + (-5) \cdot \left(\frac{26}{36}\right).$$

This sums to approximately –\$.4889. Dividing by the \$5 wager gives a house advantage of about 9.78%.

Example 5.4. Suppose that the No Lose Free Roll had no cap on winnings— so long as the shooter kept rolling winning numbers, a running NLFR wager was matched. Would that tip the edge to the player?

This question asks us to balance a payoff that's headed toward infinity with a probability of winning that's approaching 0. The new expectation is

$$E = \sum_{n=1}^{\infty} \left[(10 \cdot 2^{n-1} - 5) \cdot \left(\frac{10}{36}\right)^n \cdot \left(\frac{26}{36}\right) \right] + (-5) \cdot \left(\frac{26}{36}\right).$$

The infinite series evaluates to $\dfrac{25}{8}$, giving an expectation of

$$E = \frac{25}{8} - \frac{130}{36} = -\frac{35}{72} \approx -\$0.486.$$

The corresponding HA is 9.72%, and so the edge would still rest with the casino. ∎

One might reasonably ask, given this last result, why Vegas World capped player winnings on a \$5 initial bet at \$20,000. A possible explanation is that the casino wanted to insure itself against a long run of winning tosses, which, while very unlikely, could bankrupt the casino. Twenty-five straight winning NLFR rolls, with a \$5 initial bet, would result in a prize of \$335,544,315. While this has a very low probability (about 1 chance in 81 trillion), the consequences if it happened would be catastrophic—for the casino.

5.2 2-Die Games

Beat the Dealer

The Turning Stone Casino in Verona, New York offered a game called *Beat the Dealer* in an effort to reach out to casino patrons who might have found the often highly charged atmosphere of a craps table too intimidating [109]. Beat the Dealer is the latest version of a game once known as *High Dice* [86].

Beat the Dealer is a simple game where the player and dealer each roll a pair of dice. The dealer's dice are spun in a cage. The higher roll wins, and the house takes all ties. If the player outrolls the dealer, the bet pays off at even money.

The "house wins ties" rule and the game's simplicity seem like they should give the casino a significant advantage. For each of the eleven possible rolls of the dice, the player's chance of winning (Table 5.3) is easily found.

TABLE 5.3: Beat the Dealer win probabilities

Roll	2	3	4	5	6	7	8	9	10	11	12
P(Beat the dealer)	0	$\frac{1}{36}$	$\frac{3}{36}$	$\frac{6}{36}$	$\frac{10}{36}$	$\frac{15}{36}$	$\frac{21}{36}$	$\frac{26}{36}$	$\frac{30}{36}$	$\frac{33}{36}$	$\frac{35}{36}$

The probability p of winning Beat the Dealer can be computed by evaluating the following sum:

$$p = \sum_{X=2}^{12} P(\text{Roll } X) \cdot P(\text{Beat the dealer} \,|\, \text{Roll } X)$$

where the conditional probability in the sum is found in Table 5.3. The probability of winning is then found to be $p = \frac{4}{9}$, from which we see that the probability of losing is $1 - p = \frac{5}{9}$ and the expectation is $-\$\frac{1}{9}$, a high house edge of 11.11%—as we expected before calculating it.

A novice casino visitor would be well advised to cure his or her fear of intimidation and play craps rather than this game.

Rocket 7

One way to simplify craps is to change the rules so that every bet is resolved on every roll, making it unnecessary for players to wait through a potentially long string of rolls before learning the fate of their wagers. *Rocket 7*, a game offered in Swiss casinos, takes many basic craps bets and realigns them for play in a one-roll game [28].

Rocket 7 is played with 2d6, but without Pass or Don't Pass bets. The following bets are available:

- A player may wager on a single number from 1–6. This pays off at 2–1 if one die shows that number and 4–1 if both dice do.

- "Any 7," the simple bet that the next roll will be a 7 and a lure for many unsophisticated craps players, is available here. The odds are slightly better at a Rocket 7 table: this bet pays true odds of 5–1 unless a 7 is rolled as a 1 and a 6, in which case it pays 4–1.

- Players may bet on specific ways that a sum of 7 is rolled: 1–6, 2–5, or 3–4. Each wager pays 16–1.

- Other sums may also be bet, with payoffs ranging from 16–1 for 3 and 11 down to 6–1 for the more-likely sums of 6 and 8.

- The 1 and 6 also trigger decreased payoff odds at the "Any Double" bet. Again, true 5–1 odds are paid unless double 1s or 6s are thrown. These doubles pay 4–1.

- As with 7s, it's also possible to bet on a specified double, which pays 33–1.

- "Low" and "High" bets pay even money: the Low bet pays on a total of 3–6 and the High bet wins if the roll is 8–11. A 7 is a push on both bets, while the extreme bets of 2 and 12 push on Low and High, respectively.

In eliminating the Pass and Don't Pass lines, as well as the Come and Don't Come wagers, Rocket 7 drops the best wagers for players; this is balanced by more favorable payoffs on the bets which remain.

Example 5.5. The Any 7 bet at craps has a 16.67% HA. For Rocket 7, the expectation is

$$E = (5) \cdot \frac{4}{36} + (4) \cdot \frac{2}{36} + (-1) \cdot \frac{30}{36} = -\frac{2}{36} \approx -\$.0556,$$

giving a house edge of 5.56%—only one-third of the HA on the same bet at craps. ∎

Table 5.4 compares the HAs on Rocket 7 bets to the corresponding craps bets.

TABLE 5.4: Rocket 7 and craps HAs compared

Wager	Rocket 7 HA	Craps HA
Any 7	5.56%	16.67%
⚅	5.56%	13.89%
⚀	5.56%	13.89%
⚄	5.56%	11.11%
⚁	5.56%	11.11%

Example 5.6. The High and Low bets, due to symmetry, have the same HA. The PDF for these bets is shown in Table 5.5.

TABLE 5.5: Rocket 7 High/Low wager PDF

x	$P(X = x)$
1	$\dfrac{14}{36}$
0	$\dfrac{7}{36}$
−1	$\dfrac{15}{36}$

The house advantage can be seen to be the almost-reasonable 2.78%. ∎

The single-number Rocket 7 bets have no analog on a craps table, but are present on chuck-a-luck tables (page 229), which use 3 dice. A Rocket 7 wager on 3, for example, has the expected value of

$$E = (2) \cdot \frac{10}{36} + (4) \cdot \frac{1}{36} + (-1) \cdot \frac{25}{36} = -\frac{1}{36},$$

and so these bets carry the same 2.78% HA as on High or Low.

Casino Dice

Casino Dice is an online game from the gaming company Dynamite Idea that brings some of the ideas of roulette to the dice table. As in Rocket 7, each roll of the dice at Casino Dice resolves all wagers. Two dice are rolled electronically, and players have three types of betting options:

- Bet on the total of the two dice, with payoff odds that vary with the number selected.

- Bet on a pair of consecutive numbers, as with split bets at roulette. Again, the payoff odds depend on the numbers chosen.

- Bet on Low (2–4), Middle (5–9), or Upper (10–12), similar to the dozens bets on a roulette table.

The payoffs on single- and double-number bets are given in Table 5.6. The unusual payoff odds take full advantage of the electronic nature of Casino Dice—since there is no need to deal with physical chips, no fraction is inconvenient for player or dealer. Payoffs are rounded to the nearest penny, in the casino's favor.

A bet on Low or Upper pays off at 5.8 for 1, and a Middle bet pays 1.45 for 1.

TABLE 5.6: Payoff table for Casino Dice single- and double-number bets [12]

Single-number bets	
Roll	**Payoff odds**
2 or 12	35 for 1
3 or 11	17.5 for 1
4 or 10	11.7 for 1
5 or 9	7 for 1
7	5.8 for 1
Double-number bets	
Roll	**Payoff odds**
2 and 3 or 11 and 12	11.7 for 1
3 and 4 or 10 and 11	7 for 1
4 and 5 or 9 and 10	5 for 1
5 and 6 or 8 and 9	3.9 for 1
6 and 7 or 7 and 8	3.18 for 1

Example 5.7. These bets have very low house advantages. For example, a bet on 4 or 10 has expected value

$$E = (11.7) \cdot \left(\frac{3}{36}\right) - 1 = \frac{35.1}{36} - 1 = -\frac{.9}{36} = -\$0.025,$$

giving the casino a 2.5% edge. ∎

Similarly, low house edges, from 2.50% to 3.33%, apply to other Casino Dice bets [13]. One reason for low HAs at Casino Dice is that it's an online game, without the personnel and facility expenses associated with a physical casino. Additionally, the game can offer such relatively lucrative bets due to the optional bonus wager. After every winning bet, the player is offered the chance to double his or her money by betting on the outcome of a second dice throw. The wager is simple: The 2 dice are rolled individually, and the player bets whether the second die will be higher or lower than the first. If the two dice match, the bet may be withdrawn or continued for one additional roll; if doubles are rolled again, the bet loses. With this rule tempting players with the lure of a big payday, low HAs on the main bet are affordable for the casino.

The chance of winning the bonus bet clearly depends on the value of the first die, although whether the player chooses Higher or Lower has no effect, since the wagers are symmetric. Assume that the bet is on Higher: the player is in his or her best position when the first die shows a 1, and at the worst (hoping for a tie and another roll, since he or she cannot win) when that die comes up 6. The probability of winning is then

$$\sum_{n=1}^{6} P(\text{Roll } n) \cdot P(\text{Win} \mid \text{First die shows } n) = \sum_{n=1}^{6} \frac{1}{6} \cdot \frac{6-n}{6} = \frac{15}{36}.$$

This is also the probability of losing (just evaluate the second sum from $n = 6$ down to $n = 1$, giving the sum of the same terms in the opposite order); the probability of rolling doubles and requiring a second roll to resolve the bet is then $\frac{1}{6}$.

For a \$1 win, the bonus bet—whether the player chooses Higher or Lower—has an expectation of

$$E = (1) \cdot \left(\frac{15}{36}\right) + (-1) \cdot \left(\frac{15}{36}\right) + (A) \cdot \left(\frac{1}{6}\right) = \frac{A}{6},$$

where A represents the return on this bet if it is not settled until a second roll. We see that all of the casino's edge comes from this second roll.

On the second roll, doubles is a loser for the player, so we have

$$A = (1) \cdot \frac{15}{36} + (-1) \cdot \frac{21}{36} = -\frac{1}{6}.$$

The expected value of the original bet is then

$$E = -\frac{1}{36} \approx -.0278,$$

and the HA is 2.78%. This "double or nothing" option benefits the casino when a gambler chooses it, since the player's chance of winning is less than 50%. Every time the gambler wins, the offer is repeated, giving the casino yet another chance to win back the player's money and accumulated winnings.

Diceball

Diceball borrows some of the language of baseball for its multiroll wagers. The primary Diceball wager is the even-money Run Line bet; a "run" is defined as the shooter not throwing a 7 for 4 straight tosses. The probability of winning a Run Line bet is

$$\left(\frac{5}{6}\right)^4 = \frac{625}{1296} \approx .4825,$$

and the HA is 3.55%.

Continuing the baseball metaphor, the Grand Slam bet pays off on the number of runs the shooter is able to score. The payoffs start when 3 runs are scored. This corresponds to the shooter rolling 12–15 times without rolling a 7 and then closing out the string by rolling a 7; this probability is

$$\left(\frac{5}{6}\right)^{12} \cdot \frac{1}{6} + \left(\frac{5}{6}\right)^{13} \cdot \frac{1}{6} + \left(\frac{5}{6}\right)^{14} \cdot \frac{1}{6} + \left(\frac{5}{6}\right)^{15} \cdot \frac{1}{6} \approx .0581 \approx \frac{1}{17.22}.$$

TABLE 5.7: Diceball: Grand Slam pay table [112]

Runs	Number of rolls	Payoff	Probability
3	12–15	1–1	.0581
4	16–19	5–1	.0280
5	20–23	15–1	.0135
6	24	35–1	.0126

Similar calculations give the probabilities of scoring 4 or 5 runs. These values are shown with the payoff odds in Table 5.7. In the case of 6 runs, there is no need to include the $\frac{1}{6}$ factor; if the shooter successfully holds the dice for 24 rolls without rolling a 7, the bet pays off immediately without consideration of the 25th roll. The probability of 24 straight non-7 rolls is

$$\left(\frac{5}{6}\right)^{24} \approx .0126.$$

Example 5.8. The house advantage on the Grand Slam bet is 4.69% [112]. Suppose that a controlled dice shooter claims the ability to reduce the frequency of 7s from 1 roll in 6 to 1 roll in 7. What would the HA of a Grand Slam bet on this shooter be?

Replacing $\frac{1}{6}$ by $\frac{1}{7}$ in the calculations above gives the probabilities in Table 5.8.

TABLE 5.8: Diceball: Grand Slam probabilities for a controlled shooter with $P(7) = \frac{1}{7}$

Runs	Probability
3	.0724
4	.0391
5	.0211
6	.0247

We note that the probability of losing the Grand Slam bet, even when the shooter is this skilled, is .8427. The expected value of a $1 Grand Slam bet is then $.6070, so the bettor holds a 60.7% advantage over the casino. ∎

While there is some controversy over whether or not dice controllers have the ability to affect the dice that they claim, a dice shooter with this level of skill could win considerable money at Diceball, provided that his or her funds did not run out during the 84% of the time that the Grand Slam bet loses.

Example 5.9. If you'd like to root for 7s instead of against them, Diceball offers a "Seven Out" bet that wins if the next roll is a 7. This bet comes in two versions; casinos may choose which one to offer.

- A 4–1 payoff on any roll of 7. This is the craps "Big Red" bet, and carries the same 16.67% HA.

- A payoff of 4–1 only if 7 is rolled as ⚀ ⚅; the bet pays 5–1 on a roll of ⚁ ⚄ or ⚂ ⚃, the same payoffs offered in Rocket 7. The increased payoff makes this bet more favorable to the player; the expected value, as computed in Example 5.5, is –\$.0556, and the 5.56% HA is one-third the edge on Big Red.

∎

Street Dice

In 2014, the Downtown Grand Casino in Las Vegas began offering a new 2-die game called *Street Dice* [26]. In addition to offering different rules than standard craps, the game was novel in that it was played on the sidewalk outside the building, with dice that were about twice the size of standard casino dice and the largest dice allowed under Nevada gaming regulations. Like Die Rich, the rules for Street Dice included a three-roll limit after a player made a point on the come-out roll. In addition to keeping bets from riding too long before being resolved, this rule also limited any impact that a controlled shooter might have on the game.

As in craps, Street Dice begins with a come-out roll, which is called the "Set the Point" roll. An 11 on the first roll is an automatic winner and pays off at even money, while 2, 3, and 12 automatically lose. An initial roll of 7 is a push, rather than a win as at craps. Any other number becomes the point, and the player has three chances to re-roll that number before rolling a 7.

The payoff for a successful shooter depends both on the point and on how quickly it's re-rolled. Table 5.9 contains the payoff schedule.

TABLE 5.9: Street Dice payoffs [26]

Point	1st roll	2nd roll	3rd roll
4 or 10	5–1	4–1	3–1
5 or 9	4–1	3–1	2–1
6 or 8	3–1	2–1	1–1

Example 5.10. Let p be the probability of rolling a given point number X. What is the probability of winning a Street Craps bet with X as the point?

First, X must be rolled on the Set the Point roll and established as the point. Once this is done, the chance of re-rolling the number on the first subsequent roll is again p. To go to a second roll, it is necessary for the first roll to be neither X nor 7, and the probability of this is $1 - \frac{1}{6} - p = \frac{5}{6} - p$. The chance of then making the point on the second roll is $\left(\frac{5}{6}-p\right)\cdot p$. Repeating this analysis for the third roll gives a total probability of rerolling X as

$$p+\left(\frac{5}{6}-p\right)\cdot p+\left(\frac{5}{6}-p\right)^2\cdot p=\frac{p\cdot(36p^2-96p+91)}{36}.$$

It follows that the probability $P(X)$ of establishing X as a point and then successfully rerolling it is

$$P(X)=p\cdot\frac{p\cdot(36p^2-96p+91)}{36}=p^4-\frac{16}{6}p^3+\frac{91}{36}p^2.$$

■

For the various point values, we have the probabilities shown in Table 5.10. Taken together with the probability of $\frac{1}{18}$ of rolling an 11 on the come-out

TABLE 5.10: Street Dice win probabilities

Point X	p	$P(X)$
4 or 10	$\dfrac{1}{12}$	$\dfrac{37}{2304}\approx.0161$
5 or 9	$\dfrac{1}{9}$	$\dfrac{727}{26,244}\approx.0277$
6 or 8	$\dfrac{5}{36}$	$\dfrac{70,525}{1,679,616}\approx.0420$

roll and winning immediately, the total probability of winning a Street Dice bet is

$$\frac{1}{18}+2\cdot\left(\frac{37}{2304}+\frac{727}{26,244}+\frac{70,525}{1,679,616}\right)=\frac{95,341}{419,904}\approx.2271.$$

This low probability of winning means that the payoff odds in Table 5.9 have a gap to fill to make this an attractive game.

Example 5.11. What is the house edge on the basic Street Dice bet?

We return to the individual probabilities of rolling the point and then rerolling it within three rolls, which are found in Table 5.11.

The probability of losing a Street Dice bet with a point of X involves three disjoint events:

- The point is established on the Set the Point throw and a 7 is rolled on the next throw.

- The point is set, the second roll is neither X nor 7 and so fails to resolve the bet, and the third roll is a 7.

TABLE 5.11: Street Dice: Winning probabilities across 3 rolls

First roll	p^2
Second roll	$p^2 \cdot \left(\dfrac{5}{6} - p \right)$
Third roll	$p^2 \cdot \left(\dfrac{5}{6} - p \right)^2$

- The point is set, the next two rolls are neither X nor 7, and the final roll is anything other than X.

The probability of establishing a point X and then losing is

$$P(\text{Lose on } X) = p \cdot \left[\frac{1}{6} + \left(\frac{5}{6} - p \right) \frac{1}{6} + \left(\frac{5}{6} - p \right)^2 \cdot (1 - p) \right]$$

$$= p - \frac{91}{36} p^2 + \frac{8}{3} p^3 - p^4.$$

Denote this expression by $P(p)$.

Since the payoffs (to 1) are in a nice arithmetic progression: $\{y, y-1, y-2\}$, we can write a single equation for the expectation of a \$1 bet given that the point is X:

$$E(p, y) = y \cdot p^2 + (y - 1) \cdot p^2 \cdot \left(\frac{5}{6} - p \right) + (y - 2) \cdot p^2 \cdot \left(\frac{5}{6} - p \right)^2 - 1 \cdot P(p)$$

$$= -p + \frac{11}{36} p^2 + \frac{5}{3} p^3 - p^4 + \frac{91}{36} p^2 y - \frac{8}{3} p^3 y + p^4 y.$$

To simplify this expression further, we observe that y as defined above is a function of p; the relationship $y = 8 - 36p$ connects the two. Making this substitution gives the following formula for E in terms of p alone:

$$E(p) = -p + \frac{739}{36} p^2 - \frac{332}{3} p^3 + 103p^4 - 36p^5.$$

Collecting the three possible values for p gives Table 5.12: the expectations of the various point numbers.

Once a point is established, Table 5.12 shows that Street Dice is nearly an even game; most of the house edge comes when the game is lost, with probability $\frac{1}{9}$, on the initial roll. The final expectation on a \$1 bet is

$$E = \frac{1}{18} - \frac{1}{9} + 2 \cdot \left[E \left(\frac{1}{12} \right) + E \left(\frac{1}{9} \right) + E \left(\frac{5}{36} \right) \right] \approx -\$.0502,$$

TABLE 5.12: Street Dice: Expected values of point numbers

Point X	p	$E(p)$
4 or 10	$\frac{1}{12}$	0
5 or 9	$\frac{1}{9}$	5.6013×10^{-3}
6 or 8	$\frac{5}{36}$	-2.9352×10^{-3}

for a house advantage of about 5.02%, just under that for American roulette. ∎

The layout for Street Dice is quite simple in comparison to a craps layout. In particular, there is no "Don't Pass" option—one cannot bet against the shooter—and only one side bet, the "Brick Bet." This bet pays off if an even point number—4, 6, 8, or 10—is established on the Set the Point roll and then hits as doubles in the subsequent three rolls before a 7 is rolled. The Brick Bet must be made before a point is set; it loses immediately if the first roll is 2, 3, 5, 9, or 12 and pushes on an initial 7 or 11. If the bet hits, the payoff is 25–1.

Example 5.12. As we know, side bets such as the Brick Bet tend to have very high house advantages. Is that the case here?

The Brick Bet loses immediately if the first roll is 2, 3, 5, 9, or 12, so the probability of an instant loss is $\frac{1}{3}$. The odds of winning the Brick Bet depend on the even point number X made on that roll. Let p be the probability of rolling X. Keeping in mind that the probability of rolling X as doubles is a fixed $\frac{1}{36}$, the probability of winning is a function P of p:

$$P(p) = p \cdot \left[\frac{1}{36} + \left(\frac{5}{6} - p\right) \cdot \frac{1}{36} + \left(\frac{5}{6} - p\right)^2 \cdot \frac{1}{36}\right] = \frac{p}{1296}\left(91 - 96p + 36p^2\right),$$

and the probability of establishing and losing the Brick Bet on a point X with probability p is

$$Q(p) = p\left[\left(\frac{5}{36} + p\right) + \left(\frac{27}{36}\right)\left(\frac{5}{36} + p\right) + \left(\frac{27}{36}\right)^2 \cdot \frac{35}{36}\right] = \frac{7p}{576}(65 + 144p).$$

The expectation of the Brick Bet is

$$E = (25) \cdot \left[\underbrace{2P\left(\frac{3}{36}\right)}_{X=4,10} + \underbrace{2P\left(\frac{5}{36}\right)}_{X=6,8} \right] + (-1) \cdot \left[\frac{1}{3} + 2Q\left(\frac{3}{36}\right) + 2Q\left(\frac{5}{36}\right) \right].$$

If $p = \frac{3}{36}$, we have $P(p) \approx .0054$ and $Q(p) \approx .0780$; the corresponding values for $p = \frac{5}{36}$ are $P(p) \approx .0084$ and $Q(p) \approx .1305$. Adding everything up gives $E \approx -\$.0627$, for a house advantage of about 6.27%—fairly reasonable for a side bet. ∎

Street Dice failed to catch on with gamblers; the game lasted less than a year at the Downtown Grand.

Barbööt

Barbööt was a dice game that pitted player directly against player, with the casino acting simply as host. It was available for a time at the Aladdin Casino in Las Vegas (now Planet Hollywood).

Barbööt was based on a street game called *barbudi*, a Canadian game also favored in large northern U.S. cities [88]. The Aladdin's version of the game involved two players directly: the First Man and the Second Man. Other participants did not throw the dice, but were permitted to make side bets among themselves.

The First Man began a round by throwing two dice. If the roll was a 6–6, 6–5, 5–5, or 3–3 (5 possible rolls), he won. If the roll was a 1–1, 1–2, 2–2, or 4–4 (also 5 possible rolls), he lost. All other rolls—called neutral rolls—were disregarded; if the First Man rolled any of the other 26 combinations, the dice passed to the Second Man, who rolled with the same possibilities. The dice passed back and forth until one player rolled a winning combination. If the First Man lost by rolling doubles, the Second Man moved up and became the First Man, and the first gambler in the queue behind the Second Man became the new Second Man.

The probability that a single game of barbööt is resolved on the first roll, whether a win or a loss for the First Man, is

$$\frac{10}{36} = \frac{5}{18}.$$

The probability that a game runs for 2 rolls is simply the probability that the First Man throws a neutral roll followed by the chance that the Second Man throws a decisive roll:

$$\frac{26}{36} \cdot \frac{5}{18} = \frac{65}{324}.$$

In general, an n-roll game of bárbööt occurs when $n-1$ consecutive neutral rolls are followed by a decisive roll; this has probability

$$\left(\frac{26}{36}\right)^{n-1} \cdot \frac{5}{18} = \frac{5 \cdot 13^{n-1}}{18^n}.$$

Although a single game of barbööt could conceivably extend through many rolls, an average game lasts for only

$$\sum_{n=1}^{\infty} n \cdot \frac{5 \cdot 13^{n-1}}{18^n} = \frac{18}{5} = 3.6 \text{ rolls.}$$

Example 5.13. The Aladdin's barbööt brochure advertised that neither player holds an advantage at the game. We can confirm this by computing the First Man's chance of winning. The chance of the First Man winning on his own turn is

$$\frac{5}{36} + \left(\frac{26}{36}\right)^2 \cdot \frac{5}{36} + \left(\frac{26}{36}\right)^4 \cdot \frac{5}{36} + \cdots,$$

or

$$\frac{5}{36} \sum_{k=0}^{\infty} \left(\frac{26}{36}\right)^{2k}.$$

The chance of the First Man winning due to the Second Man throwing a losing combination is

$$\frac{26}{36} \cdot \frac{5}{36} + \left(\frac{26}{36}\right)^3 \cdot \frac{5}{36} + \left(\frac{26}{36}\right)^5 \cdot \frac{5}{36} + \cdots = \frac{5}{36} \sum_{k=0}^{\infty} \left(\frac{26}{36}\right)^{2k+1}.$$

The probability that the First Man wins is the sum of these infinite series:

$$P(\text{First Man wins}) = \frac{5}{36} \sum_{k=0}^{\infty} \left(\frac{26}{36}\right)^{2k} + \frac{5}{36} \sum_{k=0}^{\infty} \left(\frac{26}{36}\right)^{2k+1}.$$

Since the first sum includes only even powers of $\frac{26}{36}$ and the second contains only odd powers, we can combine them into a single sum:

$$P(\text{First Man wins}) = \frac{5}{36} \sum_{j=0}^{\infty} \left(\frac{26}{36}\right)^j = \frac{5}{36} \cdot S,$$

where S denotes the sum of this convergent geometric series. We have

$$S = \frac{1}{1 - \frac{26}{36}} = \frac{36}{10},$$

so

$$P(\text{First Man wins}) = \frac{5}{36} \cdot S = \frac{1}{2}.$$

It follows that the probability that the First Man loses is also ½. Since both shooters have a win probability of ½, this is indeed a fair game. ∎

In a game with no advantage, the Aladdin made its money by charging winning players a 2% commission on their winnings, or, equivalently, a 1% commission on the amount put at risk by both players. This same commission was charged to any other players making their own side bets on the game; the fee could be interpreted as a charge to players for the casino facilitating and supervising the action, much like the rake taken out of each pot in a casino's poker room.

A similar game called *barbooth* was available at Bob Stupak's Vegas World. The rules for barbooth matched those of barbööt, but gamblers placed their bets with the casino on one shooter or the other, similar to baccarat. Vegas World took its commission, 5%, as a charge applied to all bets. This was accomplished by paying off bets at 20 to 21 odds: players risked $21 but were paid only $20 if they won [6].

Survival Dice

Dean DiLullo, who would go on to a long career in casino management, invented *Survival Dice* when working at the Rio Casino in Las Vegas [81]. Survival Dice was described as "The Ultimate 2-Roll Dice Game" and "The Ultimate 3-Roll Dice Game" when it debuted in the mid-1990s. The game was based on the sum of either 2 or 3 rolls of 2d6, thus ensuring a quick resolution to the main Survival bet. Other one-roll bets were also available and resolved just as quickly.

In both games, the challenge for the shooter is to avoid two "walls," which represent losing sums. Since the walls are defined by sums of consecutive rolls, order matters when considering the Survival bet.

Example 5.14. The 3-roll game has a wall at 10 and 11. The three-roll combination 8, 9, and 3 totals 20 if rolled in that order and results in a push. If the 3 and 8 are rolled first, in either order, the cumulative sum is 11 and all Survival bets lose. There is no need to roll the dice a third time. ∎

2 Rolls

The sum of two rolls of 2d6 ranges from 4 to 24. The first wall in the 2-roll version of Survival Dice covers the sums of 4, 5, and 6. If either sum is reached in 1 or 2 rolls, all Survival bets lose. The probability of hitting the first wall on the first roll is

$$p_1 = \frac{3}{36} + \frac{4}{36} + \frac{5}{36} = \frac{12}{36} = \frac{1}{3},$$

and the probability of hitting the first wall in exactly two rolls is

$$p_2 = \underbrace{\left(\frac{1}{36}\right)^2}_{P(4)} + \underbrace{\left(2 \cdot \frac{1}{36} \cdot \frac{2}{36}\right)}_{P(5)} + \underbrace{\frac{1}{36} \cdot \frac{3}{36} + \left(\frac{2}{36}\right)^2}_{P(6)} = \frac{12}{1296}.$$

The probability of losing the Survival bet by hitting the first wall is then

$$p_1 + p_2 = \frac{1}{3} + \frac{12}{1296} = \frac{444}{1296} \approx .3426.$$

The second wall contains only the sum of 11. The probability of rolling a total of 11 in one or two rolls is

$$p_3 = \frac{2}{36} + \left(2 \cdot \frac{1}{36} \cdot \frac{4}{36} + 2 \cdot \frac{2}{36} \cdot \frac{5}{36} + \frac{6}{36} \cdot \frac{3}{36} \right) = \frac{118}{1296} \approx .0910.$$

In this expression, $\frac{2}{36}$ is the probability of rolling an 11 on the first roll. $2 \cdot \frac{1}{36} \cdot \frac{4}{36}$ covers the case of rolling a 2 and a 9 to make 11, in either order, and $2 \cdot \frac{2}{36} \cdot \frac{5}{36}$ is the probability of rolling a 3 followed by an 8 or an 8 followed by a 3. If the 11 is formed by a 7 and a 4, the 7 must have been rolled first, or else the game would have ended at the first wall with an initial 4, so there is no factor of 2 with $\frac{6}{36} \cdot \frac{3}{36}$. Finally, there is no term for 11 as a sum of 5 and 6, for an initial roll of 5 or 6 also ends the game at the first wall. The total probability of losing a Survival bet is then

$$p_1 + p_2 + p_3 = \frac{1}{3} + \frac{12}{1296} + \frac{118}{1296} = \frac{562}{1296} \approx .4336,$$

less than 50%.

What sounds like a lucrative opportunity for the gambler is revealed not to be so by a look at the pay table, Table 5.13. Final totals of 7–10 and 12–15 push, and the bet only pays off on a total of 16 or higher.

TABLE 5.13: Two-roll Survival Dice: Survival Bet pay table [18]

Total	Payoff odds
4–6	Lose: First wall
7–10	Push
11	Lose: Last wall
12–15	Push
16–19	1–1
20	2–1
21	3–1
22	4–1
23	5–1
24	100–1

The probability of winning the top prize is easily seen to be

$$\left(\frac{1}{36} \right)^2 = \frac{1}{1296},$$

the probability of rolling two straight 12s. This contributes approximately 7.7¢ to the expected value of a $1 Survival bet.

Example 5.15. Construct the probability distribution for the outcome of a $1 Survival bet if the first roll is a 7.

Starting with a 7, the final sum will range from 9 to 19, so the possible outcomes are push, win $1, and lose $1. Table 5.14 shows the second rolls that lead to each outcome and the probabilities.

TABLE 5.14: Two-roll Survival Dice: PDF for Survival bet following a 7 on the first roll

Outcome	Second rolls	Probability
Push	2,3,5,6,7,8	$\dfrac{23}{36}$
–$1	4	$\dfrac{3}{36}$
+$1	9,10,11,12	$\dfrac{10}{36}$

The corresponding expected value at this point is

$$E = (1) \cdot \frac{10}{36} + (-1) \cdot \frac{3}{36} = \frac{7}{36} \approx \$.1944.$$

■

Both versions of Survival Dice offered one-roll side bets. Among them was the Natural 12 bet that paid 32 for 1 if the next roll was ⚅ . This is a better wager than the corresponding craps bet, which pays only 31 for 1 on a next-roll 12.

Example 5.16. In the 2-roll game, the smallest paying sum is 16, so small initial rolls are deadly. The Low Roll Insurance bet offered players a chance to protect a Survival bet against low rolls. It paid 3 for 1 (2–1) on a sum of 2–5.

The expected value of this bet is

$$E = (3) \cdot \frac{1+2+3+4}{36} - 1 = -\$\frac{1}{6} \approx -\$.1667,$$

and the house held a 16.67% advantage. Once again, the reassuring name "Insurance" masked a bet with a very high HA.   ■

A High Roll Insurance bet was also present on the layout. It paid off at 3 for 1 if the next roll was 9 or greater, and had the same HA as Low Roll Insurance.

3 Rolls

The layout for 2-roll Survival Dice could fit on a standard blackjack table. 3-roll Survival Dice uses a somewhat longer layout, more easily fitted to a craps table. The two walls in this version cover rolls of 10, 11, 21, 22, and 23. Since the most likely sum of 2d6 is 7, it follows that the most common sum of 3 independent rolls is 21, and this is a wall number that wins all Survival bets for the casino.

Example 5.17. Find the probability of hitting the first wall.

Rolling a sum of 10 or 11 in 3 or fewer rolls is possible by rolling one of the combinations in Table 5.15.

TABLE 5.15: Three-roll Survival Dice: Ways to roll 10 or 11

10	
Number of rolls	**Combinations**
1	10
2	2–8, 3–7, 4–6, 5–5
3	2–2–6, 2–3–5, 2–4–4, 3–3–4
11	
Number of rolls	**Combinations**
1	11
2	2–9, 3–8, 4–7, 5–6
3	2–2–7, 2–3–6, 2–4–5, 3–3–5, 3–4–4

Each 2- or 3-roll combination can rearrange the order of the rolled numbers without excluding any possibilities, since no sums less than 10 are part of the wall. Adding everything up gives

$$P(10) = \frac{6894}{46,656} \approx .1478$$

and

$$P(11) = \frac{6588}{46,656} \approx .1412,$$

so the probability of hitting the first wall is approximately .2890. ∎

The complete pay table for the 3-roll game is shown in Table 5.16.

Example 5.18. Find the probability of rolling a total of 34 and scoring a 10–1 payout.

To reach a sum of 34, it is necessary to roll large sums and avoid both walls. The only ordered 3-roll combination that will reach 34 without intermediate sums of 10, 11, 21, 22, or 23 is 12–12–10. This has a probability of

$$\frac{1}{36} \cdot \frac{1}{36} \cdot \frac{3}{36} = \frac{3}{46,656}.$$

∎

TABLE 5.16: Three-roll Survival Dice: Survival Bet pay table [18]

Total	Payoff odds
6	100–1
7	10–1
8	5–1
9	2–1
10–11	Lose: First wall
12–20	Push
21–23	Lose: Last wall
24	Push
25	1–1
26	2–1
27	3–1
28	4–1
29	5–1
30	6–1
31	7–1
32	8–1
33	9–1
34	10–1
35	500–1
36	5000–1

Pyramid Dice

There are 21 different ways that 2 indistinguishable six-sided dice can fall when rolled. These are shown in Figure 5.4.

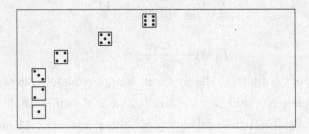

FIGURE 5.4: The 21 possible combinations of 2d6.

The challenge of rolling all 21 combinations without repeating one was at the center of *Pyramid Dice*, a game combining the action of craps with many of the betting options of roulette and a large, though improbable, jackpot. Part of the motivation for Pyramid Dice inventor Stephen Kal was to develop a dice game that might be seen as less intimidating and less complicated than craps, so every wager other than the jackpot bet was resolved on the next roll, just as in Rocket 7 or Casino Dice [16].

Pyramid Dice takes its name from a pyramid, similar to Figure 5.4, raised at the end of the dice pit. This showed the 21 possible ways for the dice to land. As each combination was rolled, it could be lit up to show the shooter's progress toward winning the jackpot. A Jackpot wager pushed if the shooter rolled a string of 7 straight different combinations and paid off on an increasing scale, shown in Table 5.17, if 8 or more different combinations were rolled consecutively. The top prize was left to the operator's discretion and could range from $100,000 to $1,000,000.

TABLE 5.17: Pyramid Dice pay table [75]

String length	Payoff odds
7	Push
8	1–1
9	2–1
10	3–1
11	4–1
12	7–1
13	9–1
14	20–1
15	50–1
16	100–1
17	500–1
18	2500–1
19	25,000–1
20	50,000–1
21	Jackpot

What is the chance of winning that jackpot? The probability of extending a string of consecutive different combinations clearly decreases as the string lengthens, but how much it drops depends on the composition of the set of combinations yet to be rolled. If there are many doubles yet to be rolled, the probability of extending the string is smaller than if the doubles have all been thrown. Of course, the chance of throwing a string with many doubles early on is smaller than the chance of starting out a string with many non-doubles.

We balance these factors by looking at one possible string: If D denotes a roll of doubles and N a roll of non-doubles, one possible jackpot string takes the form

$$N D N N N D N N N D N N N N N D N D D N N.$$

Working from left to right, the probability that the first roll is not doubles is $\frac{30}{36}$. The probability that the next roll is doubles is $\frac{6}{36}$, and the probability that the third roll is a different non-double roll from the first roll is $\frac{28}{36}$.

Continuing, the probability of successfully rolling this string, without regard for which combinations are rolled where except for the distinction between doubles and non-doubles, is

$$\frac{30 \cdot 6 \cdot 28 \cdot 26 \cdot 24 \cdot 5 \cdot 22 \cdot 20 \cdot 18 \cdot 4 \cdot 16 \cdot 14 \cdot 12 \cdot 10 \cdot 8 \cdot 3 \cdot 6 \cdot 2 \cdot 1 \cdot 4 \cdot 2}{36^{21}}.$$

If we consider rearranging the sequence of 6 Ds and 15 Ns above, we see that no matter how the terms are arranged, the denominator will always be 36^{21} and the numerator will contain the same factors as this fraction, rearranged: all of the integers from 1–6 and all of the even integers from 2–30. The first six factors come together to form 6!; the remaining terms, the product of the even integers from 30 down to 2, form the *double factorial* of 30.

Definition 5.1. Let n be a positive even integer. The *double factorial* of n, denoted $n!!$, is the product of all even integers from n down to 2:

$$n!! = n \cdot (n-2) \cdot (n-4) \cdots 4 \cdot 2.$$

$0!! = 1$, by definition.

If n is a positive odd integer, $n!!$ is the product of all odd integers from n down to 1:

$$n!! = n \cdot (n-2) \cdot (n-4) \cdots 3 \cdot 1.$$

Example 5.19. For small n, $n!!$ is easily computed by direct multiplication.

$$4!! = 4 \cdot 2 = 8.$$
$$7!! = 7 \cdot 5 \cdot 3 \cdot 1 = 105.$$
$$12!! = 12 \cdot 10 \cdot 8 \cdot 6 \cdot 4 \cdot 2 = 46,080.$$

In the current problem, we are interested in $30!! = 42,849,873,690,624,000.$ ∎

This notation allows us to write the probability of this string of Ns and Ds compactly as

$$\frac{6! \cdot 30!!}{36^{21}} \approx 6.411 \times 10^{-14}.$$

Any string consisting of 6 Ds and 15 Ns will have this same probability. There are $\binom{21}{6} = 54,264$ different ways to choose 6 positions in a string of 21 to hold Ds while the rest hold Ns. It follows that the probability of rolling 21 consecutive different combinations is

$$\binom{21}{6} \cdot \frac{6! \cdot 30!!}{36^{21}} \approx 3.479 \times 10^{-9} \approx \frac{1}{287,447,589}$$

—and thus that this event is wildly underpaid even with a million-dollar jackpot.

Example 5.20. The Instant Repeat bet is a simple wager, paying 18–1, that the first two rolls of the dice will be the same combination. This bet may only be made at the beginning of a new shooter's turn [47].

The probability of winning Instant Repeat must incorporate the possibility that the first roll is doubles. If the first roll is doubles, the chance that the second roll matches it is $\frac{1}{36}$. If not, the probability of a second-roll match is $\frac{2}{36}$. Taken together, the probability of winning is

$$\frac{6}{36} \cdot \frac{1}{36} + \frac{30}{36} \cdot \frac{2}{36} = \frac{11}{216}.$$

The corresponding expected value is

$$E = (18) \cdot \frac{11}{216} + (-1) \cdot \frac{205}{216} = -\frac{7}{216} \approx -\$.0324,$$

giving the casino a 3.24% edge. ∎

It is critical for the casino that this bet only be placed before the first roll, since if the player has the choice to make the bet after seeing the first roll, and the first roll is not doubles, the player holds a 5.56% advantage over the casino.

Pyramid Dice borrows from roulette by coloring 10 of the 21 rolls red, 10 black, and one, the ⚁ ⚂, white. These are used to facilitate color bets, with ⚁⚂ playing the role of a green zero. A bet on Red or Black pays even money, as in roulette, and a bet on White pays 15–1. While there are 21 combinations that matter in Pyramid Dice, calculations involving these bets use a denominator of 36. Since each of the Red and Black sets of rolls includes 3 doubles, the probability of each is $\frac{17}{36}$. The probability of White is $\frac{2}{36}$, and so the expected value of a Red or Black bet is

$$E = (1) \cdot \frac{17}{36} + (-1) \cdot \frac{19}{36} = -\frac{1}{18} \approx -\$.0556,$$

which is not far removed from the expectation of –$.0526 on an even-money American roulette wager. A bet on White has twice the (negative) expected value, or –$.1111.

Dice Duel

Another game in the New Games Lab at the Crown Casino is *Dice Duel*: a game challenging players to pick which of two dice, red or gold, will show the higher result when rolled [17]. Dice Duel is conducted like baccarat in that the player with the largest bet on each color is permitted to shake that die in an enclosed shaker and pass it back to the dealer, who reveals it.

The simplest bet is a bet on Gold or Red, which wins at even money if the chosen die is higher than the other. Stated like this, these two bets appear to carry no house advantage; the casino takes its cut from how ties are resolved.

If both dice roll the same number from 2–6, Gold and Red bets push, but if both dice show a 1, these bets lose. We have the probability distribution shown in Table 5.18.

TABLE 5.18: Dice Duel: PDF for the Red or Gold bet

x	$P(X = x)$
1	$\dfrac{15}{36}$
0	$\dfrac{5}{36}$
−1	$\dfrac{16}{36}$

This PDF shows that the Red and Gold bets each have a HA of 2.78%.

Dice Duel shares with craps a relatively low HA on the main bet; a craps or Dice Duel table can make more money for the casino by offering other wagers with higher edges. Dice Duel offers two other bets; the first is a Tie bet paying off at 4–1 if both dice show the same number other than 6 and 8–1 on a pair of 6s. This bet has an expected value of

$$E = (4) \cdot \frac{5}{36} + (8) \cdot \frac{1}{36} + (-1) \cdot \frac{30}{36} = -\$\frac{2}{36} \approx -\$.0556,$$

which yields a HA of 5.56%, double that of Red or Gold.

Players may also wager on an individual number from 1–6, and are paid 2–1 if one die shows their number, 4–1 if both do. This bet has the same expectation as Red or Gold:

$$E = (2) \cdot \frac{10}{36} + (4) \cdot \frac{1}{36} + (-1) \cdot \frac{25}{36} = -\$\frac{1}{36} \approx -\$.0278,$$

so Dice Duel is perhaps more similar to baccarat than craps in that every bet except Tie has a relatively low HA.

Two-Up/Heads & Tails

In its original form, the Australian game *Two-Up* is found in legal and illegal gaming establishments [39]. Current Australian law provides that, in most cities, the game is only legal on Anzac Day, April 25, as a tribute to war veterans [33]. Main Street Station in downtown Las Vegas brought Two-Up to America.

Two-Up is played with 2 coins, which are tossed by a designated player, the *spinner*, from a small flat piece of wood called the *kip*. The two coins can land both Heads, both Tails, or one of each: a throw termed "Odds." The street version of Two-Up is a fair game: players bet with each other on whether the spinner will toss Heads or Tails; if the coins land Odds up, all bets carry over to the next throw.

In a casino version of the game, the spinner wagers that he or she will toss Heads three times in a row before tossing either Tails once or 5 consecutive Odds. The three tosses of Heads need not be consecutive; if this were a requirement, a single round of Two-Up could last a long time. This bet pays 7½–1.

Example 5.21. The longest possible string of tosses without resolving Two-Up wagers is the 14-toss string OOOOHOOOOHOOOO. This string has probability

$$\left(\frac{1}{2}\right)^4 \cdot \frac{1}{4} \cdot \left(\frac{1}{2}\right)^4 \cdot \frac{1}{4} \cdot \left(\frac{1}{2}\right)^4 = \frac{1}{65,536}.$$

Regardless of the outcome of the 15th toss, the bet will be settled.

An average game of Two-Up consists of 3.2 tosses. ∎

Casino Two-Up offers three wagers to other players [39].

- Players may bet on Heads or Tails on each toss. If Odds is thrown, all bets are frozen until Heads or Tails is tossed, or until Odds occurs 5 straight times. Heads and Tails are even-money wagers, and both lose on 5 tosses of Odds.

- A third option is to bet that the spinner will toss Odds 5 times in a row. This pays 30–1. The expected value of this bet is

$$E = (30) \cdot \left(\frac{1}{2}\right)^5 + (-1) \cdot \frac{31}{32} = -\frac{1}{32} = -\$.03125.$$

To compute the HA on the Heads and Tails bets, we note that the probability of losing either one of these bets is $\frac{1}{2} + \frac{1}{32}$, the sum of the probabilities of tossing the other pair and tossing 5 consecutive Odds. The HA can then be seen to be the same $3\frac{1}{8}\%$ as the Odds bet.

Several Las Vegas casinos offered a version of Two-Up called *Heads & Tails* with special six-sided dice replacing the coins. These dice, a pair of which are shown in Figure 5.5, bear an H on three sides and a T on the other three sides. As when tossing a coin, the probability of rolling a head or tail on these dice is ½.

At Caesars Palace, the game was played at a standard blackjack or baccarat table, and the dice were shaken in an enclosed container. Two versions of Heads & Tails were available. The Slow Lane offered two even-money bets: on Heads and Tails. If the dice came up Odds, all bets pushed. In the Fast Lane, players could bet on Heads, Tails, or Odds. A winning Odds bet paid even money, while Heads and Tails bets paid 3–1 when they won.

Both the Slow Lane and Fast Lane paid bets at true odds; Caesars Palace charged a 5% commission on all winning Fast Lane wagers, and so derived its edge there.

FIGURE 5.5: Heads & Tails dice from the El Cortez Casino, Las Vegas.

Example 5.22. Alternately, a casino could establish its advantage at Heads & Tails by paying off winning bets at less than true odds. One proposal [67] suggests that Heads and Tails bets pay 2–1 and Odds pay 4–6. The corresponding HA on Heads and Tails wagers is then 25%—perhaps too high. Odd bets carry a $16\frac{2}{3}\%$ advantage, which also runs high.

An alternate pay table offers 5–2 payoff odds on Heads and Tails and 4–5 for Odds. Here, the HAs would be 12.5% and 10% respectively, which is more player-friendly without calling for unusual fractional payoffs. ■

5.3 3-Die Games

Banca Francesa

Banca Francesa (Portuguese, "French Bank") is a 3-die game played in some casinos in Portugal. Three dice are rolled by the dealer, and three wagers are available to players [39]:

- The *Small* bet, which pays off at even money if the sum of the dice is 5, 6, or 7.

- The *Big* bet, which pays off at even money of the sum is 14, 15, or 16.

- The *Aces* bet, which pays off at 60–1 if all three dice show a 1.

Since there are only three possible bets, the Banca Francesa layout (Figure 5.6) is quite simple, consisting of two concentric circles with a sector cut out to accommodate the Aces bet. The dealer rolls the dice near the top of this layout.

Players may bet entirely within the Big and Small regions, where chips count their full value, or may place their chips on the outside edge of those regions, where they are counted at half their face value. Aces bets must always go at full value, and so are bet completely within that section of the layout.

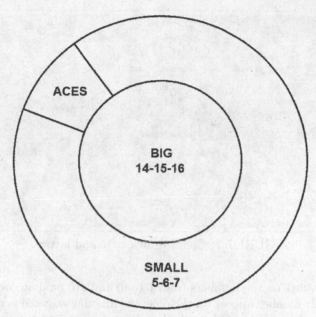

FIGURE 5.6: Banca Francesa layout [39]

There are 6 ways to roll a sum of either 5 or 16, 10 ways to roll a 6 or 15, 15 ways to roll a 7 or 14, and 1 way to roll a 3, so it follows that only 63 of the possible 216 ways for three dice to fall correspond to combinations that resolve the three bets on offer. This leaves 153 neutral rolls which are ignored, and the dice then rerolled.

Since only 63 rolls affect the outcome of the game, the expected value of a $1 bet on either Small or Big can be determined using 63 as the denominator:

$$E = (1) \cdot \left(\frac{31}{63}\right) + (-1) \cdot \left(\frac{32}{63}\right) = -\$\frac{1}{63} \approx -\$0.0159.$$

For the Aces bet, the expected value is

$$E = (60) \cdot \left(\frac{1}{63}\right) + (-10) \cdot \left(\frac{62}{63}\right) = -\$\frac{2}{63} \approx -\$0.0317.$$

It follows that the bets at Banca Francesa have quite reasonable house advantages: 1.59% for Big and Small, and 3.17% for Aces.

Chuck-a-Luck

Chuck-a-Luck is a simple game played with three dice. In its original form, it is not often seen in casinos, but can still be found in carnivals and, in expanded form, in the casino game sic bo (page 231). The three dice are spun in a wire cage shaped like an hourglass, as shown in Figure 5.7.

FIGURE 5.7: Chuck-a-luck cage and layout.

Gamblers bet on the numbers from 1 to 6, and are paid according to how many of their number appear on the dice: the amount wagered is matched for each die showing the selected number, so the payoff is 1 to 1 if one die shows the number, 2 to 1 on 2 matches, and 3 to 1 if all three match.

Example 5.23. Let X be the number of ⚃s that appear when the cage is spun. Since there are six sides on each die, there are $6 \cdot 6 \cdot 6 = 216$ ways that the dice can land. We note that the numbers showing on the three dice are independent, and as a result, X is a binomial random variable with parameters $n = 3$ and $p = \frac{1}{6}$. The probability distribution function for X is

$$P(X = x) = \binom{3}{x} \cdot \left(\frac{1}{6}\right)^x \cdot \left(\frac{5}{6}\right)^{3-x}.$$

∎

The number 4, of course, can be replaced by any number from 1–6 without changing the PDF. This function is tabulated in Table 5.19.

It's then easy to compute the expected return:

$$E = (1) \cdot \frac{75}{216} + (2) \cdot \frac{15}{216} + (3) \cdot \frac{1}{216} + (-1) \cdot \frac{125}{216} = -\frac{17}{216} \approx -\$.0787,$$

which shows that the house has a 7.87% edge—par for the course for a simple wager such as this. Rocket 7 offers the same bets, but gamblers face a lower HA there, even with fewer dice in play.

This simple game can be improved for gamblers by increasing the top payoff. At the sic bo tables at the Crown Casino, the payoff for matching all three dice is 12–1 rather than 3–1. While this sounds like it might be a windfall opportunity for players, the low probability of hitting a triple means that the HA on this enhanced wager only falls to 3.70%.

TABLE 5.19: Chuck-a-Luck PDF

x	$P(X = x)$
0	$\dfrac{125}{216}$
1	$\dfrac{75}{216}$
2	$\dfrac{15}{216}$
3	$\dfrac{1}{216}$

Sic Bo

Sic bo is a 3-dice-in-a-cage game that extends chuck-a-luck. The sic bo layout depicted in Figure 5.8 offers a wide variety of bets based on the fall of three dice. In the bottom two rows one finds all of the chuck-a-luck bets: the next-to-last row contains betting spaces and the last row shows the corresponding payout for bets in this row.

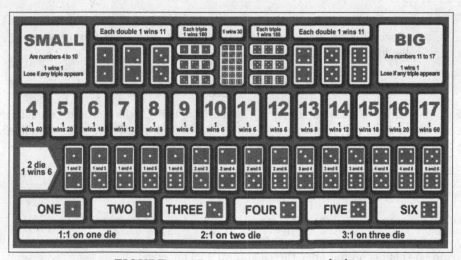

FIGURE 5.8: Sic bo betting layout. [99]

Example 5.24. A new category of sic bo bets is the six *double* bets at the top of Figure 5.8. These bets pay off at 11–1 if two or three dice show the selected number.

The probability of winning a specified double bet is

$$\binom{3}{2} \cdot \left(\frac{1}{6}\right)^2 \cdot \left(\frac{5}{6}\right) + \frac{1}{216} = \frac{16}{216}.$$

In this equation, the first term in the sum on the left computes the probability of rolling exactly two of the chosen number and the second is the probability of rolling three of a kind.

The HA on this bet is $\frac{24}{216}$, or 11.11%.   ∎

Figure 5.8, with its 11–1 payoff on a double bet, is not the only sic bo pay table in use. At the Horseshoe Casino in Hammond, Indiana, a double wager pays only 8–1, which causes the HA to jump to 33.33%.

Example 5.25. The fifteen bets on two different numbers appearing on the three dice, which pay at 6–1, are sometimes called *domino* bets for the resemblance of the two dice to a domino. The chance of winning a domino bet is found by enumerating all possible cases; we shall consider the ⊡⊡ bet as an example.

- If the third die is neither 1 nor 2, there are $_3P_3 = 6$ ways to roll that 3-die combination, for a total of 24 winning rolls.

- If the third die is 1 or 2, there are 3 ways to choose the unpaired die, and so 3 winning combinations for 1-1-2 or 1-2-2. This gives 6 more winning rolls, for a grand total of 30.

The probability of winning is then $\frac{30}{216}$, or slightly less than 1 in 7.   ∎

A possible additional sic bo bet is a *Straight* bet, which would pay off if the 3-dice form a sequence: 1-2-3, 2-3-4, 3-4-5, or 4-5-6. There are 4 possible straights, and each one allows $_3P_3 = 6$ ways to assign numbers to dice. This product gives 24 winning rolls, so the probability of winning is $\frac{24}{216}$.

If the Straight bet pays off at 7–1, its expectation is

$$E = (7) \cdot \frac{24}{216} + (-1) \cdot \frac{192}{216} = -\frac{24}{216} = -\$\frac{1}{9},$$

and so this bet would offer the casino an 11.11% edge.

Rolling in Sequence

A proposed alteration of sic bo rolled the dice individually and permitted players to continue placing bets after the first and second dice were thrown [4]. Since additional knowledge can change probabilities, the payoffs on later bets must change. For example, the Any Triple bet, which pays 30–1 at sic bo any time the three dice show the same number, pays only 4–1 here. That being so, it would be foolish to make this bet before two dice have been thrown and have turned up the same number, when the probability of winning has risen from $\frac{1}{36}$ to $\frac{1}{6}$.

Sic bo's Any Triple bet carries a HA of 13.89%. When made under the terms detailed above, Any Triple's expected value is

$$E = (4) \cdot \frac{1}{6} + (-1) \cdot \frac{5}{6} = -\$\frac{1}{6},$$

and so the HA has increased to 16.67%. What seems like a favor to the players has turned into an improved game for the casino. If Any Triple was paid at $4\frac{1}{6} - 1$, the HA would remain 13.89%.

Example 5.26. A wager that makes use of this format is the *Total* bet. This bet can be made before the first, second, or third dice have been rolled, and pays 9–2 if the player chooses the correct sum of the 3 dice.

The best way to make a Total bet is to wait until after the first die has been rolled. If the number on that die is X, bet on $X + 7$. This gives a win probability of $\frac{1}{6}$.

Suppose that the first two dice have a sum of Y. The HA of a Total bet at this point can be computed by imagining simultaneous \$2 Total bets on the 6 numbers $Y + 1$ through $Y + 6$—this guarantees one winning bet and 5 losing bets. Since this gambler will win \$9 and lose \$10, the expected return is $-\$1$ and the house advantage is

$$- \left(\frac{-1}{12} \right) \approx 8.33\%.$$

∎

The *Mix* bet is based on the parity, odd or even, of each of the three dice. There are two options: even/odd/odd (EOO) and odd/even/even (OEE). All even or all odd rolls are not considered here. The order of the dice is not considered here: for EOO, the even roll can be first, second, or third, and is similar for the odd roll in OEE. Mix bets can be placed at any time before the third die is rolled, and pay even money.

Example 5.27. When is the best time to make a Mix bet?

Clearly, it would be foolish to make this bet when it cannot be won, as when making an OEE bet after the first two dice have come up odd. Before the first die is rolled, we can use the binomial distribution with $p = \frac{1}{2}$ to show that

$$P(EOO) = P(OEE) = 3 \cdot \left(\frac{1}{2} \right)^3 = \frac{3}{8}.$$

If the first die is odd, then the probabilities change:

$$P(EOO) = \frac{1}{2}$$

$$P(OEE) = \frac{1}{4},$$

whereas if the first die is even, these probabilities are reversed:

$$P(EOO) = \frac{1}{4}$$

$$P(OEE) = \frac{1}{2}.$$

After two dice are rolled, it may be true that one of the two outcomes is impossible, as was noted above. If the first two dice have opposite parities, then both EOO and OEE can still win. Any outcome which is still possible after 2 dice has probability ½.

The highest probability of winning at any point in the 3-roll sequence is therefore ½, which gives a fair wager, and so this bet is best made when it has a 50% chance of winning. Whether that is after 1 or 2 dice have been thrown is a matter of player choice, but it should be noted that even at the highest win probability, it is not possible for players to gain the edge. ∎

If the casino wishes to maintain an advantage on the Mix bet no matter when it's made, there are two common adjustments recommended by game designers:

1. *Charge a commission on winning wagers.* By charging a 5% commission, effectively paying off winning bets at 19–20 odds, the casino can lock in a positive edge. This is routine for bets on the Banker hand at baccarat.

2. *Designate one or more winning combinations as pushes or losses.* Sic bo does this with the High and Low bets, which exclude triples.

The Mix bet cannot exclude triples, since those will always be losing OOO or EEE combinations, but is there a way to exclude some rolls, possibly adjusting the even-money payoff, to produce a wager with a reasonable casino edge?

Suppose, for example, that any 3-die combination including a pair was designated a push for the EOO and OEE bets. There are 216 ways for the 3 dice to land.

- 27 of them are composed of 3 odd numbers, and 27 contain 3 even numbers.

- The 162 arrangements that remain are evenly divided, 81–81, between EOO and OEE sets.

- Of the 81 OEE sets, 54 of them do not contain paired even numbers. Similarly, 54 EOO sets do not have paired odd numbers. Either bet then includes 27 winning combinations that have been converted to pushes.

At a payoff of x–1, the expected return on a $1 Mix wager, either EOO or OEE, when pairs push is

$$E = (x) \cdot \frac{54}{216} + (0) \cdot \frac{27}{216} + (-1) \cdot \frac{135}{216} = \frac{54x - 135}{216}.$$

The winning payoff on a Mix bet could be increased to 2–1, and the casino would then have a 12.5% edge.

This is the situation when a Mix bet is made before any dice are rolled. If the first two numbers are one odd and one even, we have the following probabilities for an OEE bet:

$$P(\text{Win}) = \frac{1}{3}, \ P(\text{Push}) = \frac{1}{6}, \ P(\text{Lose}) = \frac{1}{2}.$$

At 2–1 for a win, the expected value is

$$E = \frac{2}{3} - \frac{1}{2} = \frac{1}{6},$$

and the player has the advantage. Changing the payoff to 3–2 makes this a fair bet when made after an EO or OE combination has been rolled.

The commission option may be preferable here, as it gives a more reasonable HA even as it slows down game play while commissions are computed and collected.

Tai Sai

Tai Sai ("big-small") is an extension of sic bo that may be found at a number of casinos in Asia and Oceania. In addition to the full array of sic bo bets, tai sai players may bet in several different ways on the numbers rolled on the three dice. Recall that sic bo allows players to place a domino bet on a pair of numbers and collect a 6–1 payoff if those two numbers are both rolled. Tai sai gamblers can extend that wager to a bet on all three numbers rolled as a triple of numbers, rather than as a sum. The payoffs on these bets depend on whether or not the triple contains a pair [106]:

- If the three numbers are all different, the payoff is 30–1. There are $_3P_3 = 6$ ways to win this bet, corresponding to the 6 ways to assign three numbers to three distinct dice.

- If the trio includes a pair, there are only 3 ways to win, as the only choice is which of the three dice bears the unpaired number. The payoff is 60–1.

As the payoff odds on the first bet are half that of the second while the probability of winning is doubled, these two bets have the same house edge. We find for the first bet that

$$E = (30) \cdot \frac{6}{216} + (-1) \cdot \frac{210}{216} = -\frac{30}{216} = -\frac{5}{36}.$$

The house edge on either bet is then 13.89%. Just because a new bet is available doesn't mean that it's a good choice.

Example 5.28. A second group of extra tai sai bets allows players to bet on a group of four numbers from 1–6. These bets pay off at 7–1 if any three of the four numbers are rolled. Four of the $\binom{6}{4} = 15$ possible tetrads are available for wagering: 1–2–3–4, 2–3–4–5, 2–3–5–6, and 3–4–5–6. The probability of a four-number combination bet winning is

$$\binom{4}{3} \cdot {}_3P_3 \cdot \left(\frac{1}{6}\right)^3 = 4 \cdot 6 \cdot \left(\frac{1}{6}\right)^3 = \frac{24}{216} = \frac{1}{9},$$

and the corresponding expectation is

$$E = (7) \cdot \frac{1}{9} + (-1) \cdot \frac{8}{9} = -\$\frac{1}{9}.$$

The house advantage is 11.11%—still not worth considering.   ∎

How does a four-number bet compare to four separate three-number bets covering all $\binom{4}{3} = 4$ choices of three of the four numbers? Neither is a good bet, but the collection of four simultaneous three-number bets carries the same 13.89% HA as each bet does individually, so a four-number bet is slightly less awful for the gambler.

Yee Hah Hi

Yee hah hi is another variation of sic bo, one that uses three six-sided dice with pictures on each side rather than numbers [52]. This game is common in Macau and has made some recent inroads into Australian casinos. Yee hah hi's name translates into English as "fish shrimp crab," and these are three of the pictures on the dice. Three yee hah hi dice are shown in Figure 5.9.

FIGURE 5.9: Yee hah hi dice. From left to right, the coin, crab, and rooster faces are uppermost.

Some yee hah hi dice include numbers along with the pictures, and each picture appears in one of three colors: red, green, or blue. The images are associated with numbers (when present) and colors as shown in Table 5.20.

TABLE 5.20: Yee hah hi dice [52]

Picture	Number	Color	Picture	Number	Color
Fish	1	Red	Coin	4	Blue
Shrimp	2	Green	Crab	5	Green
Gourd	3	Blue	Rooster	6	Red

With or without the numbers, yee hah hi is largely identical to sic bo; what is different about this game is an additional color wagering option, which is implemented by assigning one of three colors to each picture. Available color wagers at yee hah hi are these:

- *Color Triple* bets, which pay off at 23–1 if all three dice show the same selected color.

- *Any Color Triple*, which pays 7–1 if all three dice are the same color.

- *Color Double* bets, which pay 3–1 if two of the three dice show the selected color. This bet loses if all three dice show the chosen color.

- *Color Single*, which pays 1–1 if *exactly one* die shows the selected color.

The Color Single bet pays no bonus if more than one die shows the color chosen by the player—indeed, the bet loses if more than one die turns up the selected color. Can the casino maintain an edge while paying Color Single bets like in chuck-a-luck or sic bo, with a payoff based on how many dice show the right color?

We begin by constructing a probability distribution for X, the number of dice showing a given color, in Table 5.21. For color bets, the dice are effectively 3-sided.

This table tells us all we need to know. As constructed, the Color Single bet has a win probability of $\frac{12}{27}$. If we allow payoffs on 2 or 3 occurrences of the chosen color, the probability of winning rises to $\frac{19}{27}$—well over .5, and thus no set of payoffs greater than or equal to 1–1 will give the casino an edge.

Similar analysis based on Table 5.21 will show that including the case where all three dice are the same color as a winning combination in the Color Double bet is a losing proposition for the casino.

Example 5.29. Find the HA on an Any Color Triple bet.

Since this bet covers any of the three colors, we need not consider the color of the first die in this analysis; all that matters is that the other two dice

TABLE 5.21: Yee hah hi: Color probability distribution for a selected color

x	$P(X{=}x)$
0	$\left(\dfrac{2}{3}\right)^3 = \dfrac{8}{27}$
1	$3 \cdot \dfrac{1}{3} \cdot \left(\dfrac{2}{3}\right)^2 = \dfrac{12}{27}$
2	$3 \cdot \dfrac{2}{3} \cdot \left(\dfrac{1}{3}\right)^2 = \dfrac{6}{27}$
3	$\left(\dfrac{1}{3}\right)^3 = \dfrac{1}{27}$

match it. The probability that all three dice show the same color is

$$p = \left(\frac{1}{3}\right)^2 = \frac{1}{9}.$$

The expected value of a $1 bet is then

$$E = (7) \cdot \frac{1}{9} + (-1) \cdot \frac{8}{9} = -\$\frac{1}{9} \approx -\$0.1111,$$

and so the HA is 11.11%.   ∎

By repeating the calculations in Example 5.29 for the other color bets, we find that all four of these bets have the same 11.11% HA, and thus that there are far better yee hah hi bets among the standard sic bo wagers.

Super Color Sic Bo

The evolution of three-dice-in-a-cage carnival games from chuck-a-luck through sic bo and tai sai may appear to have reached its limit with yee hah hi. To do anything more with this game would seem to necessitate a bigger change than an expanded bet field or more features on 3d6.

The developers of *Super Color Sic Bo* rose to the challenge.

Super Color Sic Bo is a game operated at the Indonesian online casino Royal369 which uses three 12-sided dice, or dodecahedra, with pentagonal faces bearing the numbers from 1–12 in three different colors [105].

- **Red**: 1, 4, 9, and 12.

- **Blue**: 2, 5, 8, and 11.

- **Green**: 3, 6, 7, and 10.

Instead of a cage, the dice are tossed inside a glass dome by an electromechanical device.

Fourteen types of wager are available on the layout. Chuck-a-luck wagers on a single number are not present, perhaps because the expected value of such a bet with chuck-a-luck payoffs is

$$E = (1) \cdot \frac{363}{1728} + (2) \cdot \frac{33}{1728} + (3) \cdot \frac{1}{1728} + (-1) \cdot \frac{1331}{1728} \approx -\$.5203,$$

and a wager with a 52% HA is too high to be attractive without the lure of a massive payoff—something much higher than a mere 3–1.

Example 5.30. The *Double Digits* bet comes in two types. The first, which pays 11-4, is a bet that 2 of the 3 dice show the same number, any number. The probability of winning a Double Digits bet when a pair is rolled is

$$\binom{12}{2} \cdot \binom{2}{1} \cdot \binom{3}{2} \cdot \left(\frac{1}{12}\right)^3 = \frac{11}{48}.$$

In this formula, the factor $\binom{12}{2} = 66$ counts the number of ways to choose the two numbers that appear on the dice. $\binom{2}{1} = 2$ determines the number of ways to choose which of those two numbers appears on 2 dice, and $\binom{3}{2} = 3$ is how many ways we may select which two dice show that number. Once the arrangements are so counted, we know which number must appear on each die. The probability of rolling the required number on each of the three dice is $\frac{1}{12}$.

Additionally, this bet can win if the three dice show the same number; this has probability

$$12 \cdot \left(\frac{1}{12}\right)^3 = \frac{1}{12^2} = \frac{1}{144}.$$

Taken together, the probability of winning a Double Digits bet is $\frac{17}{72}$, and the expected value of a \$4 bet is

$$E = (11) \cdot \left(\frac{17}{72}\right) + (-4) \cdot \left(\frac{55}{72}\right) \approx -\$.4583.$$

The HA is then 11.46%.

11–4 seems like an unusual payoff. Why was this selected instead of something simpler such as 3–1 or 5–2?

This question can be answered by looking at the HA of the bet using these other payoffs. At 3–1 (12–4), the payoff on a \$4 bet increases to \$12, and that \$4 bet has expectation

$$E = (12) \cdot \left(\frac{17}{72}\right) + (-4) \cdot \left(\frac{55}{72}\right) \approx -\$.2222,$$

which reduces the HA to 5.56%, which game designers considered too low. If the bet pays 5–2 (equivalent to 10–4), we have

$$E = (10) \cdot \left(\frac{17}{72}\right) + (-4) \cdot \left(\frac{55}{72}\right) \approx -\$.6944,$$

and the 17.36% HA might be too high to attract players.

A specific double number bet, paying 45–1, may be made by specifying the number to appear at least twice. The probability of winning this bet is

$$\binom{3}{2} \cdot \left(\frac{1}{12}\right)^2 \left(\frac{11}{12}\right) + \left(\frac{1}{12}\right)^3 = \frac{34}{1728} \approx .0197,$$

and the house advantage on a \$1 bet is 9.49%.   ∎

Example 5.31. Triples allow only one betting option. A bet that the three dice will show the same number, any number from 1–12, pays 130–1. The probability of rolling triples is

$$12 \cdot \left(\frac{1}{12}\right)^3 = \frac{1}{144},$$

and so the HA on this bet is 9.03%.   ∎

The different colors on Super Color Sic Bo dice numbers are involved in a collection of color-specific bets. The *Three Stars* bet pays 3–1 if all three colors are represented on the three dice. Though the order of the dice does not matter, counting the winning die rolls is facilitated by imposing an informal order on the dice. The first die may be any color. Once that color is set, the probability that the second die has a different color is $\frac{2}{3}$, and then the chance that the third die bears the remaining color is $\frac{1}{3}$. Multiplying gives the probability of winning a Three Stars bet as $\frac{2}{9}$, leading to an 11.11% HA—right in line with the other bets we have examined.

As with the Double Digits bets, one may make two different bets that the dice will all show the same color. A bet on a specified color pays 23–1 and has a win probability of

$$\left(\frac{1}{3}\right)^3 = \frac{1}{27}.$$

Alternately, a simple bet that the three dice will be the same color pays 7–1, with corresponding probability found by adding the probabilities of all-red, all-blue, and all-green rolls to yield

$$3 \cdot \frac{1}{27} = \frac{1}{9}.$$

Both bets have the same HA: 11.11%.

The Straight bet in sic bo (page 232) was a hypothetical wager. This bet is available in Super Color Sic Bo, where it pays off at 25–1 if the three dice are in sequence, from 1–2–3 up through 10–11–12. The "around-the-corner" straights 11–12–1 and 12–1–2 are specifically excluded as winning combinations. There are 10 different three-number straights; the numbers in each can be arranged among the three dice in 3! = 6 ways, which gives 60 three-die rolls out of 1728 that yield a straight. This wager has an expected value of

$$E = (25) \cdot \frac{60}{1728} + (-1) \cdot \frac{1668}{1728} = -\frac{7}{72} \approx -\$.0972.$$

If the around-the-corner straights were counted as winning rolls, the expected value would rise to \$.0833—a positive value that would give the player an $8\frac{1}{3}\%$ advantage.

There are 1728 different ways for the three 12-sided dice to land. Of these, 865 have an odd sum and 863 have an even sum, which is close enough to a 50/50 split to justify *Even* and *Odd* even-money bets. With no further rules beyond "win if the sum is the parity that you bet on," the HA on these bets would be too low for casino interest—indeed, the Odd bet would carry a .116% player edge, exactly the amount of the HA on Even. To guarantee a meaningful house advantage, the Odd bet pushes if the sum is 13 and the Even bet pushes on a sum of 26. There are 66 ways to roll a 13 and 65 ways to roll a 26; the probability distributions for these bets are shown in Table 5.22.

TABLE 5.22: Super Color Sic Bo Odd/Even probability distributions

	Odd	Even
$P(\textbf{Win})$	.4621	.4615
$P(\textbf{Push})$	.0382	.0376
$P(\textbf{Lose})$	.4997	.5009

The HAs on these bets are considerably lower than other Super Color Sic Bo wagers: 3.76% for Odd and 3.93% for Even.

Super Color Sic Bo also offers a collection of bets linked to animals of the Chinese zodiac. The *Zodiac* bets are based on the sum of the three dice, as shown in Table 5.23.

These bets are symmetrically arranged about the mean of 19.5 for the sum of a roll of three 12-sided dice, so there are 6 mathematically different bets in the table. The bets' HAs are, in general, in line with other three-dice-in-a-cage games; the highest probability of winning comes on the Snake and Horse bets, while the lowest HAs, 15.74%, are found on the Rat and Pig wagers. Table 5.24 shows the win probability and HA of each bet, with mathematically identical bets listed together.

TABLE 5.23: Super Color Sic Bo Zodiac bets [105]

Animal	Sums	Payoff Odds
Rat	3–8	26–1
Water buffalo	9–11	13–1
Tiger	12–13	11–1
Rabbit	14–15	8–1
Dragon	16–17	7–1
Snake	18–19	6–1
Horse	20–21	6–1
Goat	22–23	7–1
Monkey	24–25	8–1
Chicken	26–27	11–1
Dog	28–30	13–1
Pig	31–36	26–1

TABLE 5.24: Super Color Sic Bo Zodiac payout information

Animals	P(Win)	House Advantage
Rat, Pig	.0324	15.74%
Water buffalo, Dog	.0631	18.00%
Tiger, Chicken	.0700	22.97%
Rabbit, Monkey	.0961	23.15%
Dragon, Goat	.1146	19.79%
Snake, Horse	.1238	25.69%

Three Dice Celo

Three Dice Celo debuted at the Foxwoods Casino in Mashantucket, Connecticut. This is a casino version of a 3-die street game called 4–5–6 that was popular in Alaska [98]. The game incorporates aspects of craps, baccarat, and sic bo.

Gamblers may bet on Banker or Player, as in baccarat. Play begins with the dealer throwing 3 dice for the Banker side. The following rolls are immediate winners:

- 4–5–6.

- Any triple.

- Any pair where the third die is a 6, except for 1–1–6, on which all bets push.

These rolls are immediate losers:

- 1–2–3.

- Any pair where the third die is a 1, except for 1–1–1.

If the dice show a pair, the third die, if not a 1 or 6, establishes the Banker's point. The dealer rolls the dice until a winning combination, losing combination, 1–1–6, or point is thrown. The probability of an immediate win is

$$\frac{6}{216} + \frac{6}{216} + \frac{12}{216} = \frac{24}{216},$$

while the probability of an immediate loss is

$$\frac{6}{216} + \frac{15}{216} = \frac{21}{216}$$

and the probability of establishing a point is

$$\frac{\binom{3}{2} \cdot 5 \cdot 4}{216} = \frac{60}{216}.$$

Taken together with the probability of all bets pushing on an initial roll of 1–1–6, which is $\frac{3}{216}$, we see that 108, or 50%, of all rolls either resolve the bets or establish a point. The probability of establishing a point is equally distributed across the four possible points 2, 3, 4, and 5, so the probability of making a particular point is $\frac{15}{216}$.

If the Banker establishes a point, the dice pass to one of the gamblers, who rolls for the Player side. Winning and losing rolls are the same for the Player. If the Player side establishes a point, Player wins if the point is higher than the Banker's point and the Banker wins if its point is higher. If both sides make the same point, all bets push.

In computing the win probabilities, we concern ourselves only with the 108 rolls that resolve one side of the game, so the denominators change from 216 to 108. The probability of a Banker win is

$$\frac{26}{108} + \frac{60}{108} \cdot \frac{21}{108} + \frac{15}{108} \cdot \left[\frac{15 + 30 + 45}{108}\right] = \frac{301}{648} \approx .4645.$$

The probability of a push is just the probability of an initial 1–1–6 plus the probability of both sides establishing the same point:

$$\frac{3}{216} + 4 \cdot \left(\frac{15}{108}\right)^2 = \frac{59}{648} \approx .0910.$$

It follows that the probability of a Player win is

$$1 - \frac{301}{648} - \frac{59}{648} = \frac{4}{9} \approx .4444.$$

If we disregard push decisions, the probability of the Banker winning a resolved game is

$$\frac{.4645}{.4645 + .4444} \approx .5110.$$

Since this is greater than ½, winning Banker bets are paid at .95–1, with a 5% commission deducted. The resulting HA is a low .35%, while the Player bet pays even money and carries a 2.01% HA.

As in craps, the low HAs on the main bets are supported by an array of additional bets with higher casino edges.

Example 5.32. Three Dice Celo offers a one-roll "Fast Field" bet similar to the craps Field bet. This bet pays even money if the dice sum to 5, 7, 9, 11, or 13, 2–1 on a 17, and 3–1 on a 3.

The casino's edge in Fast Field comes from the fact that a roll of 15 is a losing roll, the only odd sum that loses. The probability of rolling a 15 is $\frac{10}{216}$. Without the bonus payouts on 3 and 17, the expected value of a Fast Field bet would be

$$E = (1) \cdot \frac{98}{216} + (-1) \cdot \frac{118}{216} = -\frac{20}{216} \approx -\$.0926.$$

This gives a HA of nearly 10%, which is high enough to allow game designers to give something back with the bonuses. The expected value of a $1 Fast Field bet is then

$$E = (1) \cdot \frac{94}{216} + (2) \cdot \frac{3}{216} + (3) \cdot \frac{1}{216} + (-1) \cdot \frac{118}{216} = -\frac{15}{216} \approx -\$.0694,$$

and the resulting HA is just under 7%. ∎

There are two different types of bets on triples available in the center of the Three Dice Celo layout. The one-roll version pays 180–1 if a specific named triple is rolled on the next roll; an "Any Triple" roll pays 30–1 if the three dice all show the same number. These bets are identical to triple bets in sic bo. The named triple bet has an expected value of

$$E = (180) \cdot \frac{1}{216} + (-1) \cdot \frac{215}{216} = -\frac{35}{216} \approx -\$.1620$$

and Any Triple's expected value is

$$E = (30) \cdot \frac{6}{216} + (-1) \cdot \frac{210}{216} = -\frac{30}{216} \approx -\$.1389,$$

which is slightly higher, though still not a lucrative bet.

Players may also choose the "Decision Roll" option when betting on triples, which pays 90–1 if a triple is specified and 15–1 on "Any Triple," but neither wins nor loses if the roll is one of the 108 inconclusive rolls. This restricts the universal set for the bets to 108 rolls. Since the payoffs have been cut in half while the probabilities have doubled, these bets have the same expectations and HAs as the one-roll versions.

Example 5.33. A different kind of Three Dice Celo bet is "Beat the 2," which wins if a decision roll does better than establish 2 as a point by rolling 4–5–6, rolling any triple, or establishing 3, 4, or 5 as the point. The probability of winning this bet is

$$\frac{6}{108} + \frac{6}{108} + 3 \cdot \frac{15}{108} = \frac{57}{108} > \frac{1}{2},$$

which is addressed in game design by setting the payoff at 1–2, less than even money. It follows that the expected value of a \$2 bet is

$$E = (1) \cdot \frac{57}{108} + (-2) \cdot \frac{51}{108} = -\frac{45}{108} \approx -\$.4167,$$

which gives the casino a 20.83% edge. The 1–2 payoff tips things back in the casino's favor, perhaps too far.

Suppose that Beat the 2 pays off at x to 1, where x is necessarily less than 1 in order to give the casino an edge. We have the general expectation

$$E(x) = (x) \cdot \frac{57}{108} + (-1) \cdot \frac{51}{108} = \frac{57x - 51}{108}.$$

Table 5.25 shows a selection of payoff odds and their corresponding HAs. There is room to make this bet more favorable to players while preserving a

TABLE 5.25: Three Dice Celo: Payoff options for the Beat the 2 bet

Payoff odds	x	HA
2–3	.6667	12.04%
3–4	.7500	7.64%
4–5	.8000	5.00%
5–6	.8333	3.24%

reasonable house advantage. ∎

Beat the 2 is a fair bet if $57x - 51 = 0$, which would yield payoff odds of 17–19. In addition to providing the casino no advantage, implementing this payoff would make the game more difficult to run due to the less-than-round numbers.

Fan Tan and Fan Tan Astic

Fan Tan is a game played with buttons that is found primarily in Asian casinos including the Lucky Ruby Border Casino and Lucky-89 Casino, both in southern Cambodia. The game has a long history in China, dating back to at least 645 BCE [1]. After a long underground run in Chinese communities across America, fan tan was first licensed in the USA in 1967, when it debuted at the New China Club in Reno.

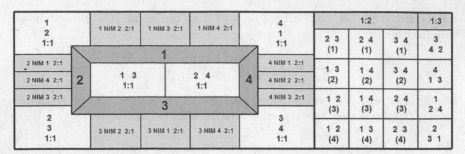

FIGURE 5.10: Fan tan layout.

Though the betting layout, shown in Figure 5.10, looks very complicated, fan tan is quite simple. The dealer extracts a small pile of buttons (approximately 60) from a larger pile and counts them out by fours. Players bet on the number of buttons left as the last group of four is removed: 1, 2, 3, or 4. Wagers may be made on one, two, or three of the four numbers:

- **Fan** is a bet on one number to win. Fan bets are made in the darker gray single-numbered trapezoids of Figure 5.10, and pay off at 3–1.

- **Nim** is a bet on two numbers: one to win and the other to push. The bet pays 2–1 if the winning number comes up and pushes if the other number does. These are made in the lighter gray sections to the left of the layout, with the winning number listed first, before the word "Nim." "3 Nim 1," for example, is a Nim bet on 3 to win and 1 to push.

- **Kwok** bets on two numbers, and pays off at 1–1 if either one wins. These wagers are placed in the corner spaces surrounding the Fan rectangle at the left, or in the centermost rectangle.

- **Nga Tan** designates two numbers to win and one to push. The payoff on a winning number is 1–2. These 12 bets are located in the column headed 1:2 to the right. The number bet to push is shown here in parentheses.

- **Sheh-Sam-Hong** is a three-number bet, all to win. The payoff is 1–3, and the four combinations are shown to the right, under the 1:3 column heading.

Some fan tan layouts are equipped with lights that illuminate all winning wagers when the dealer pushes a button signaling the result, as with some sic bo layouts.

Under the entirely reasonable assumption that all four outcomes are equally likely, each of these bets is fair as written. The house charges a 5% commission on winning bets, which accounts for the casino edge. No commission is charged on bets that push. The effective payoffs on the five bets listed above are then 2.85–1, 1.9–1, .95–1, .475–1, and .317–1.

Example 5.34. The corresponding expectation on a Fan bet is then

$$E = (2.85) \cdot \frac{1}{4} + (-1) \cdot \frac{3}{4} = -\frac{.15}{4} = -\$.0375,$$

and the house advantage is 3.75%. ■

Example 5.35. Since no commission is charged on pushes, the expectation of a Nga Tan bet is

$$E = (.475) \cdot \frac{1}{2} + (0) \cdot \frac{1}{4} + (-1) \cdot \frac{1}{4} = -\frac{.05}{4} = -\$0.0125,$$

giving a lower HA of 1.25%. ■

For such a simple game, the house edges on fan tan bets are surprisingly small.

Fan Tan Astic, a product of Xcite Gaming, is a variation of fan tan played with dice. In Fan Tan Astic, Fan Tan's buttons are replaced by three dice, and the players bet on the remainder when the sum of the three dice is divided by 4 [24]. This speeds up the game; rolling dice and adding is much quicker than counting out buttons by 4s. There is, of course, some loss of suspense.

- A bet on 0 pays off if the sum of the dice is 4, 8, 12, or 16.

- A bet on 1 pays off if the sum of the dice is 5, 9, 13, or 17.

- A bet on 2 pays off if the sum of the dice is 6, 10, 14, or 18.

- A bet on 3 pays off if the sum of the dice is 3, 7, 11, or 15.

Rolling three dice means that there are 216 different combinations. The gambler looking for an edge might consider a chart of all possible sums, which shows that 55 different sums leave a remainder of 0 or 1 when divided by 4, while only 53 combinations reduce to 2 or 3. The four number bets are not starting out on equal footing here.

These four standard Fan Tan Astic bets all pay 3–1. Rather than charging a commission as in fan tan, the house takes its edge by declaring a push on all winning bets whenever a pair of 6s is rolled. This takes away 3 winning combinations from a bet on 0 (6-6-4, 6-4-6, and 4-6-6), 2, or 3; and removes 6 winning combinations from a bet on 1. Triple 6s are not counted as double 6s for a bet on 2.

As partial compensation for pushes, bets on the numbers 1, 2, and 3 carry a bonus if certain triples are rolled:

- Triple 3s pay 12–1 for a bet on 1.

- Triple 5s pay 11–1 for a bet on 3.

- Triple 6s pay 11–1 for a bet on 2.

TABLE 5.26: Fan Tan Astic probability distribution for a bet on 1 [24]

Payoff x	$P(X = x)$
3	$\dfrac{48}{216}$
12	$\dfrac{1}{216}$
0	$\dfrac{6}{216}$
–1	$\dfrac{161}{216}$

Example 5.36. Find the expectation for a bet on 1.

There are four outcomes for a $1 bet: win $3, win $12, push, and lose $1. The corresponding probability distribution function is shown in Table 5.26.

The expectation is then

$$E = \frac{144 + 12 + 0 - 161}{216} = -\frac{5}{216} \approx -\$.0231.$$

The HA is 2.31%. ∎

Only small changes in the probabilities are necessary to assess a bet on 0, 2, or 3. Let this number be denoted by N and the expectation of the bet by E_N. There are four variables in the expression for E_N; define them as follows:

- w = the number of winning combinations.

- p = the number of pushing combinations.

- $y = \begin{cases} 1, & \text{if there is a triple that pays extra. } (N \neq 0) \\ 0, & \text{if no triple pays extra. } (N = 0) \end{cases}$

- x = the payoff (to 1) on a triple, where $x = 0$ if no triple pays extra.

The expectation of a bet on the number N is then a function of these four variables:

$$E_N(w,p,y,x) = (3) \cdot \frac{w}{216} + (x) \cdot \frac{y}{216} + (-1) \cdot \frac{216 - w - p - y}{216}$$

$$= \frac{4w + xy + p + y - 216}{216}.$$

When the values of $w, p, y,$ and x are plugged into this formula, we find that all four single-number bets have the same 2.31% HA that prevails on the number 1. The imbalance we noted at the beginning has been eliminated by careful design of the game to include bonuses for some, but not all, triples.

Triples also play a part in an optional Fan Tan Astic bet that pays off an escalating amount if any triple is rolled, similar to the 30–1 "Any Triple" bet at sic bo or Three Dice Celo. The payoffs on the "Any 3 of a kind" bet are collected in Table 5.27.

TABLE 5.27: Fan Tan Astic Any 3 of a kind payoffs [24]

Triple	Payoff odds
1-1-1	15–1
2-2-2	20–1
3-3-3	25–1
4-4-4	30–1
5-5-5	40–1
6-6-6	50–1

Example 5.37. How does Any 3 of a kind compare to Any Triple?

The expected value of an Any 3 of a kind bet is

$$E = (15 + 20 + 25 + 30 + 40 + 50) \cdot \frac{1}{216} + (-1) \cdot \frac{210}{216} = -\frac{30}{216},$$

while the expectation on Any Triple is

$$E = (30) \cdot \frac{6}{216} + (-1) \cdot \frac{210}{216} = -\frac{30}{216}.$$

The bets are mathematically identical, with a HA of 13.9%. ∎

Qilin

Qilin ("kee'-lin") is the name of a mythical Chinese unicorn which was adapted to a nonstandard 3-die game that was available at the Grosvenor Casinos chain in the United Kingdom beginning in 2011. The game used 3 special dice of different colors; several of the dice faces bore a "Q" instead of a number.

- **Red**: 1, 1, 3, Q, Q, Q.

- **Blue**: 1, 1, 1, 2, Q, Q.

- **Green**: 1, 2, 2, 3, 3, Q.

In a nod to the game's Chinese namesake, the numbers on the dice were displayed with 1, 2, or 3 horizontal lines in the style of Chinese numerals.

Qilin dice are enclosed in a tumbler for easy controlled tossing. Two types of bets are available.

- Players may wager on whether or not any specific die will show a Q when the dice are tossed. An additional Q bet is a wager that all three dice will show a Q; this is called a "Qilin" roll. A Qilin bet pays off at 34–1 if three Qs are thrown; bets on individual Qs pay off according to Table 5.28 if a Qilin is thrown.

TABLE 5.28: Qilin Q bet payoff odds [114]

Color	Payoff without Qilin	Payoff with Qilin
Red	1–1	Push
Blue	2–1	1–1
Green	5–1	4–1

In the absence of a Qilin, these bets pay off at true odds; the casino gets its edge from the reduced payoff when a Qilin is thrown.

Example 5.38. The probability of a Qilin is

$$\frac{3}{6} \cdot \frac{2}{6} \cdot \frac{1}{6} = \frac{1}{36},$$

where the fractions in the product arise, respectively, from the red, blue, and green dice.

A $1 Qilin bet then has expected value of

$$E = (34) \cdot \frac{1}{36} + (-1) \cdot \frac{35}{36} = -\frac{1}{36} \approx -.0278,$$

and the HA is 2.78%.

The expected value of a red Q bet is

$$E = (1) \cdot \left(\frac{3}{6} - \frac{1}{36} \right) + (0) \cdot \frac{1}{36} + (-1) \cdot \frac{3}{6} = -\frac{1}{36} \approx -.0278.$$

∎

- Additional wagers, on the sum of the dice with a Q counting as 0, are available. The pay table is shown in Table 5.29.

TABLE 5.29: Qilin Sum bet payoff odds [114]

Sum	Payoff odds
1	10–1
2	5–1
3	3–1
4	3–1
5	6–1
6	13–1
7	25–1
8	100–1

Example 5.39. Rolling a 1 on the three dice requires that two Qs be thrown. The chance of a 1 involves three cases, depending on which dice show the Qs and which one rolls a 1. In the formula for the probability,

$$\left(\frac{2}{6}\cdot\frac{2}{6}\cdot\frac{1}{6}\right)+\left(\frac{3}{6}\cdot\frac{3}{6}\cdot\frac{1}{6}\right)+\left(\frac{3}{6}\cdot\frac{2}{6}\cdot\frac{1}{6}\right)=\frac{19}{216}\approx .0880,$$

the three factors in each term in parentheses are listed again in order: red, blue, and green.

With a 10–1 payoff, the casino has an edge of 3.24%. ■

Rock/Paper/Scissors

In addition to showing people what Casino War might look like, the movie *Vegas Vacation* gave some insight into another possible casino innovation. In the same seedy-looking casino where Casino War is on offer, characters played by Chevy Chase and Randy Quaid encounter rock/paper/scissors, which is not played according to any standard rules. Here is a description of a possible version of RPS more suitable for casino play:

- Players make a bet on either Rock, Paper, or Scissors. Players may bet on more than one of the three spots, if they wish.

- The dealer rolls three distinguishable dice, perhaps in a sic bo or chuck-a-luck cage. One die is designated Rock (red, perhaps), one Paper (purple), and one Scissors (silver).

- The outcome of the roll, and the casino's choice in the game, is the die with the highest number rolled, with the following two exceptions:

  1. If two dice show the same number, then the third die is automatically the casino's outcome, even if it shows a lower number than the other two dice.

Example 5.40. If the dice come up Red 1, Purple 3, and Silver 3, then the two 3s cancel out and the casino plays Rock.   ∎

2. If all three dice show the same number, this is known as "Dynamite," and all player bets lose. (Disclaimer: Dynamite is not accepted as part of the standard RPS canon; see worldrps.com for details.)

- Bets are then paid off at even money in accordance with the standard RPS payoff formula:

    - Rock beats Scissors

    - Paper beats Rock

    - Scissors beats Paper

- If the player and casino have the same result, the bet is a push.

Assuming that Dynamite is not rolled, every round of RPS concludes with one bet that wins, one that loses, and one that pushes, giving the appearance of a balanced game. Dynamite, of course, is what gives the casino its edge. The probability of Dynamite is simply

$$6 \cdot \left(\frac{1}{6}\right)^3 = \frac{1}{36}.$$

Example 5.41. Find the probability that the dealer chooses Rock.

Rock becomes the dealer's choice under either of the following circumstances:

- All 3 dice show different numbers, with the red die the highest.

    There are $\binom{6}{3} = 20$ ways to roll 3 different numbers on 3 dice. For each way, there are $_3P_3 = 6$ ways to assign numbers to dice, giving 120 distinguishable rolls. One-third of these, or 40, have the largest number on the red die.

- The purple and silver dice show a pair, and the red die shows a different number.

    There are 6 ways to pick the pair, and then 5 ways to pick the number on the red die, for a total of 30 rolls with a pair that win for Rock.

The probability of Rock is then

$$\frac{40 + 30}{216} = \frac{70}{216} \approx .3241.$$

∎

Due to the symmetry of RPS, this is also the probability that the dice give the dealer Paper or Scissors.

If the dealer rolls Rock, players win if they bet on Paper and lose if they bet on Scissors. Similar payoffs ensue for other dealer outcomes; the players also lose if Dynamite is rolled. The house advantage for a $1 wager on Rock, Paper, or Scissors is

$$-\left[(1)\cdot\frac{70}{216}+(0)\cdot\frac{70}{216}+(-1)\cdot\frac{76}{216}\right]=\frac{1}{36}\approx2.78\%.$$

Rock/paper/scissors under these rules might be a viable casino game.

5.4 4-Die Games

Casino Bowling

Just as Diceball was a carnival dice game based on baseball, *Casino Bowling*, patented in 1996, used dice in a game of chance modeling another popular sport [8].

A round of Casino Bowling begins with the roll of 2d6. The sum of this first roll is interpreted as follows:

- **2–5**: Open frame.
- **6–9**: Roll a second pair of d6s.
- **10–12**: Strike.

If a 6 through 9 is rolled on the first toss, two additional d6 are rolled. The outcome of the round then depends on the sum of the 4 dice:

- **8–14**: Open frame.
- **15–21**: Spare.

Example 5.42. If the first roll is ⚅ ⚃, for a total of 8, the dice are rolled again. If that roll is ⚂   , the total is 14, and the round is scored as an open frame. ∎

Since strikes involve only 1 roll, the probability of a strike is simple to compute:
$$P(\text{Strike})=\frac{6}{36}=\frac{1}{6}.$$

The probability of a spare can be written using conditional probability:

$$P(\text{Spare})=P(\text{Sum of 15–21 in 2 rolls}\,|\,\text{First roll is 6–9}).$$

If the first roll is a 6, then a second roll of 9–12 will complete a spare. We have

$$P(\text{Spare} \mid \text{First roll} = 6) = P(9 - 12) = \frac{10}{36},$$

and so

$$P(\text{First roll} = 6 \text{ and Sum of second roll} \geqslant 9) = \frac{5}{36} \cdot \frac{10}{36} = \frac{50}{1296}.$$

Similarly, the following probabilities contribute to $P(\text{Spare})$:

$$P(\text{First roll} = 7 \text{ and Sum of second roll} \geqslant 8) = \frac{6}{36} \cdot \frac{15}{36} = \frac{90}{1296}.$$

$$P(\text{First roll} = 8 \text{ and Sum of second roll} \geqslant 7) = \frac{5}{36} \cdot \frac{21}{36} = \frac{105}{1296}.$$

$$P(\text{First roll} = 9 \text{ and Sum of second roll} \geqslant 6) = \frac{4}{36} \cdot \frac{26}{36} = \frac{104}{1296}.$$

Adding these 4 values gives

$$P(\text{Spare}) = \frac{349}{1296} \approx .2693.$$

A similar analysis with open frames gives

$$P(\text{Open frame in 2 rolls}) = \frac{371}{1296}.$$

To this, we must add the probability of an open frame on the first roll, which is $\frac{10}{36}$, yielding

$$P(\text{Open frame}) = \frac{731}{1296} \approx .5640.$$

With these probabilities computed, we consider the menu of available Casino Bowling wagers. Before any dice are rolled, players must make an Ante wager either on Open or on Mark, which covers both spares and strikes. The probability of winning a Mark bet is

$$P(\text{Spare}) + P(\text{Strike}) = \frac{349}{1296} + \frac{1}{6} = \frac{565}{1296}.$$

The payoff odds for Mark depend on the winning roll, and are shown in Table 5.30.

A \$1 Mark bet has an expected return of −1.70¢, so the HA on this bet is only 1.70%.

Since the probability of an open frame is greater than ½, game balance necessitates payoff odds less than 1–1 on the Open bet. Open pays off at 3–4 odds, leading to an expected value of:

$$E = (3) \cdot \frac{731}{1296} + (-4) \cdot \frac{565}{1296} = -\frac{67}{1296}.$$

The HA is 1.29%.

TABLE 5.30: Casino Bowling: Mark bet pay table [8]

Winning roll	Payoff odds	Probability
Spare	1–1	.2693
Strike with 10	1–1	.0833
Strike with 11	2–1	.0556
Strike with 12	3–1	.0278

Example 5.43. Bowlers refer to 3 strikes in a row as a "turkey." Casino Bowling offers a wager that the next 3 rounds will be strikes, which pays 200–1. Since successive rounds are independent, the probability of rolling a turkey is

$$\left(\frac{1}{6}\right)^3 = \frac{1}{216},$$

so this bet has a HA of

$$\frac{15}{216} \approx 6.94\%.$$

∎

A related wager pays 10–1 if three successive rounds are marks.

A standard game of bowling lasts 10 frames, and a perfect game, arising when 12 consecutive strikes are rolled, scores 300. Casino Bowling includes an optional progressive wager on a string of strikes, which pays off according to Table 5.31. In this table, payoffs for 8 or more strikes include a fixed aggregate payoff, escalating to $25,000,000, divided among all winners plus a percentage of an optional progressive jackpot. For example, if a wager ends after 8 strikes are rolled, every player making this bet divides $125,000 and receives 5% of the progressive pot.

TABLE 5.31: Casino Bowling: Progressive Strike bet pay table [8]

Strikes	Payoff odds
3	75–1
4	300–1
5	1250–1
6	7500–1
7	12,500–1
Strikes	**Payoff**
8	$125,000 + 5%
9	$500,000 + 10%
10	$1,000,000 + 25%
11	$10,000,000 + 50%
12	$25,000,000 + 100%

The probability of 12 strikes in a row is

$$P(12 \text{ strikes}) = \left(\frac{1}{6}\right)^{12} = \frac{1}{2,176,782,336},$$

so this bet may be undervalued unless the progressive jackpot is very large. For $n = 3$ to 11, the probability of n strikes is

$$P(n \text{ strikes}) = \left(\frac{1}{6}\right)^n \cdot \frac{5}{6} = \frac{5}{6^{n+1}}.$$

Example 5.44. The probability of losing a Progressive Strike bet is simply the chance that 0–2 strikes are thrown; this is

$$\frac{5}{6} + \frac{1}{6} \cdot \frac{5}{6} + \left(\frac{1}{6}\right)^2 \cdot \frac{5}{6} = \frac{215}{216} \approx .9954.$$

∎

When viewed that way, this is not an attractive wager. However, if that probability can be overcome, the HA on this bet is a reasonable 5.64% for a single bettor without the progressive jackpot option. If the progressive jackpot reaches $1,478,851.09, then the expectation for a single player is positive.

Yo!

Craps lore holds that it's bad luck to say the number 7 at the tables while a roll is in progress. Since the number 11 sounds so much like 7, the rolling of an 11 is frequently greeted by shouts of "Yo!," short for "yo-leven," in an effort to celebrate this winning roll while respecting the superstition. The designers of the 4-die game *Yo!* took this cheer and turned it into the name of a new carnival game, another one targeting millennial gamblers.

Yo! uses four dice, two red and two yellow. The red dice are called *game dice*; the yellow dice are known as *bonus dice* and only count in game play if they add up to the 11 in the game's name. The game is a slightly simplified version of craps with 3 main bets and 5 single-roll side bets. The three main bets are the Play, More, and Yo! bets [119].

- The **Play** wager functions like a pass line bet in craps. It wins and pays 4–1 on the first roll if the game dice show 11, pushes (rather than winning, as in craps) if the roll is 7, and loses on a 2, 3, or 12.

  Any other number on the game dice becomes the point, and the shooter rolls the dice until rerolling the point or rolling a 7. As in craps, successfully rerolling the point before rolling a 7 pays even money.

  A Yo! analog of the craps Place bet is the Late Play option. This bet may be made by any player after a point has been established, but may only be made on the current point. Since the Place bet favors the casino by removing the come-out roll, where the gambler has a momentary edge, from consideration, late bets on an established point are welcome.

- A **More** bet can only be made immediately after a point is established, and functions like the free odds bet in craps. The More bet wins if the shooter successfully makes his or her point. As in craps, the payoff on a winning More bet depends on the point: 2–1 on a 4 or 10, 3–2 if the point is 5 or 9, and 6–5 on 6 or 8.

As with free odds, the More bet carries a 0% HA.

- **Yo!** bets, like More bets, must be placed when a point is established. A Yo! bet wins when an 11 shows on the bonus dice before a 7 is rolled on the game dice. The Yo! bet pushes if the game dice seven out on the same roll. The payoff odds on Yo! depend on the simultaneous roll of the game dice, as shown in Table 5.32.

TABLE 5.32: Yo! bet pay table for an 11 rolled on the bonus dice [119]

Game dice roll	Payoff odds
11	20–1
Any pair	4–1
7	Push
Any other roll	1–1

The average payoff from a winning \$1 Yo! bet is

$$(20) \cdot \frac{2}{36} + (4) \cdot \frac{6}{36} + (0) \cdot \frac{6}{36} + (1) \cdot \frac{22}{36} = \frac{86}{36} = \$2\frac{7}{18}.$$

The chance of rolling an 11 on the bonus dice before rolling a 7 on the game dice is found by evaluating an infinite series. The probability of rolling an 11 without a 7 on a single roll is

$$p = \frac{2}{36} \cdot \frac{5}{6} = \frac{10}{216}$$

and the probability of rolling neither number on its designated dice on a single roll is

$$q = \frac{34}{36} \cdot \frac{5}{6} = \frac{170}{216}.$$

The desired probability of an 11 before a 7 is then

$$p + qp + q^2p + q^3p + \cdots = \sum_{k=0}^{\infty} p \cdot q^k = \frac{p}{1-q} = \frac{10}{46},$$

slightly less than 1 chance in 4.

With that probability of winning the Yo! bet, the expectation is

$$E = \left(2\frac{7}{18}\right) \cdot \frac{10}{46} + (-1) \cdot \frac{36}{46} = -\frac{109}{414} \approx -\$.2633.$$

Despite the exclamation point, this is not a bet to get excited about.

The 5 side bets are all one-roll bets that have action on the next roll of the game dice. These bets are arranged in a self-service arc on the layout, and can be made directly by players without assistance from dealers, since they rely only on the next roll and are not interrupting a sequence of rolls that determine the outcome of other wagers.

- The *11* bet is the same as the 11 bet on a craps table. It pays 15–1 if the next roll is an 11 and loses otherwise. Its HA is the same as the craps bet: 11.11%.

- *2,3,12* collects those 4 rolls into one wager paying 7–1.

- A *Pairs* option doesn't quite cover all of the pairs. It pays off 7–1 if a pair of 2s through 5s is rolled—double 1s and 6s lose this bet. Just like 2,3,12, this bet has 4 winning rolls and 32 losing rolls, and pays 7–1. These bets also give a 11.11% edge to the house.

- *High* and *Low* bets cover the numbers 8, 9, 10 and 4, 5, 6 respectively, and so don't quite go as high or low as is possible on 2d6. "High Middle" and "Low Middle" might be more accurate descriptions of these bets, which each pay off at 3–2.

  A High bet or Low bet has 12 winning rolls and 24 losing rolls, so payoff odds of 2–1 would result in a fair wager. With the 3–2 payoff, a $2 bet has expectation

$$E = (3) \cdot \frac{12}{36} + (-2) \cdot \frac{24}{36} = -\$\frac{12}{36} = -\$\frac{1}{3}.$$

Dividing by the $2 wager shows that the HA is 16.67%.

All of these bets lose if the game dice show a sum of 7, which makes monitoring them easier for dealers. If the shooter rolls a 7 on the game dice, the dealers can sweep the arc clear and collect all bets.

5.5   5-Die Games

Shake a Day

In Montana, a common and legal progressive game of chance played at bars is *Shake a Day*. Players contribute 50¢ to a pot of money and roll 5 standard dice. A player rolling 5 of a kind takes the entire pot—Montana state law provides that no liquor establishment may make a profit from the game, so there is no rake, or house percentage, deducted, as is the case in casino poker rooms. Other payoffs for lesser rolls such as four of a kind vary from bar to bar. Shake a Day gets its name from the rule that a customer is only allowed to roll the dice once per day, though a single establishment may run multiple pots simultaneously, and a player may pay into each pot once a day [65].

The probability of winning at Shake a Day by rolling 5 of a kind is

$$6 \cdot \left(\frac{1}{6}\right)^5 = \frac{1}{1296}.$$

The initial factor of 6 reflects the fact that any one of the 6 numbers on a die may be repeated 5 times. If there are no other prizes on offer, it follows that Shake a Day has a positive expectation when the jackpot exceeds $648, since each shake costs 50¢. However, the "one roll per day per person" rule sharply limits any effort to exploit a positive expectation by repeated play.

Example 5.45. How many daily shakes are necessary to give a 50% chance of winning the top prize?

We must solve for n in the equation

$$P(\text{Win}) = 1 - \left(\frac{1295}{1296}\right)^n = .5;$$

the solution is

$$n = \frac{\ln .5}{\ln \left(\dfrac{1295}{1296}\right)} \approx 897.97.$$

Rounding up, we find that if you play Shake a Day 898 times, investing $449, your chance of rolling 5 of a kind just exceeds ½. This represents almost 2½ years of once-daily play. ∎

The Montana lottery offers a variety of electronic games at video lottery terminals, or VLTs. The Treasure Play game suite includes a dice-based game based on Shake a Day called *$hake a Day*, which offers players a simulated roll of 5 standard dice and pays off if a full house, 4 of a kind, or 5 of a kind are rolled, as per Table 5.33. The top prize is a progressive jackpot starting at $1000.

We computed the probability of rolling 5 of a kind above. The probability of rolling 4 of a kind is

$$(6) \cdot \binom{5}{4} \cdot \left(\frac{1}{6}\right)^4 \cdot \frac{5}{6} = \frac{25}{1296},$$

and the probability of rolling a full house is

$$(6) \cdot \binom{5}{3} \cdot \left(\frac{1}{6}\right)^3 \cdot (5) \cdot \left(\frac{1}{6}\right)^2 = \frac{50}{1296}.$$

TABLE 5.33: Pay table for the Montana Lottery's $hake a Day: $3 ticket [58]

Roll	Payoff
5 of a kind	Jackpot
4 of a kind	$50
Full house	$3

While these probabilities are valid for a game played with real dice, $hake a Day is an electronic game with different programmed probabilities:

- $P(\text{5 of a kind}) = \dfrac{1}{5000}$.

- $P(\text{4 of a kind}) = \dfrac{1}{55.56} = \dfrac{18}{1000}$.

- $P(\text{Full house}) = \dfrac{1}{2.5} = \dfrac{2}{5}$.

Only the probability of a full house—which merely pays out a refund of the $3 ticket price—is greater than the probability found from considering real dice. Using these values and letting J denote the jackpot amount, we see that the expected value of a $3 $hake a Day ticket is

$$E = (J) \cdot \frac{1}{5000} + (50) \cdot \frac{18}{1000} + (3) \cdot \frac{2}{5} - 3 = \frac{J - 4500}{5000}.$$

This is positive only when the jackpot exceeds $4500, and so $hake a Day is unlikely ever to be a positive-expectation game.

Example 5.46. In Example 5.45, we saw that 898 shakes were necessary to assure a 50% chance of winning the Shake a Day jackpot. How many shakes are necessary to give a 50% chance of winning the top prize at $hake a Day?

The only thing that changes in moving from Shake a Day to $hake a Day is the probability of winning. We want to solve the equation

$$P(\text{Win}) = 1 - \left(\frac{4999}{5000}\right)^n = .5$$

for n, and here the solution is

$$n = \frac{\ln .5}{\ln \left(\dfrac{4999}{5000}\right)} \approx 3465.4.$$

If you play $hake a Day once per day, which is not a restriction in force, you will play for 9½ years and invest $10,398 before your chance of winning exceeds 50%. ∎

Nutz

Nutz is a 5-die game originating in the New Games Lab at the Crown Casino. This game shares certain similarities with the home game Yahtzee, in that players roll 5 dice up to three times in an effort to roll various combinations [66]. Part of the appeal of Nutz, as with craps, is surely that gamblers have the option to participate actively in the game by rolling the dice themselves; this is not a choice afforded blackjack or roulette players.

Bettors may choose from 7 different wagers corresponding to particular arrangements of 5 dice; any given bet wins if the precise combination selected is rolled in no more than 3 tosses. Table 5.34 shows the wagers and the payoff odds.

TABLE 5.34: Wagering options for Nutz [66]

Wager	Payoff
Nutz (5 of a kind)	100–1 on 1st roll
	25–1 on 2nd or 3rd roll
Pair	25–1
Straight or No Hand	9–1
Quads (4 of a kind)	5–1
Trips (3 of a kind)	5–1
Two Pairs	3–1
Full House	2–1

While it may seem counterintuitive that a pair pays off more highly than four of a kind, this ranking is a direct result of the rules of the game, which somewhat limit the player's choices [66]. Players may not, for example, break up a rolled four-of-a-kind to try to reduce it to a pair.

- If the first roll is a 5-of-a-kind, straight, full house, or no hand (5 different dice, not in sequence), there are no further rolls.

- If the first roll yields two or more matching dice, including two pairs, the matching dice are set aside and the nonmatching dice are rerolled.

- After the second roll, any unmatched dice are rolled a third and final time.

Example 5.47. If the first roll is ⚀⚂ ⚂⚃, all but the 3s are returned to the dice cup and rerolled. If this roll produces ⚄ , all three dice are rerolled again. If the final roll is   , the roll is two pairs, and that wager pays off at 3–1.

If the first roll is ⚀   , then the 4 will be rerolled twice unless a 1 or 2 appears on the second roll. At this point, the only possible winning bets are Two Pairs and Full House. If the player has bet on Two Pairs, he or she hopes not to complete a full house. If the second or third roll is ⚀, then the Full House bet wins. ∎

The Straight or No Hand bet can only win on the first roll, if 5 different numbers are rolled. The probability of this event is easy to compute if we consider the order of the 5 dice. We think of the 5 numbers rolled as arranged in order; this may be done in $_6P_5 = 720$ ways. The probability of each die showing its assigned number, once the sequence has been determined, is $\frac{1}{6}$, and so the probability is

$$_6P_5 \cdot \left(\frac{1}{6}\right)^5 = \frac{_6P_5}{6^5} = \frac{720}{7776} \approx .0926.$$

This bet has an expectation of

$$E = (9) \cdot \frac{720}{7776} + (-1) \cdot \frac{7056}{7776} = -\frac{576}{7776} \approx -.0741.$$

The HA on this bet is 7.41%—well within carnival bet territory.

Example 5.48. It follows from the rerolling rules that winning a Pair bet requires rolling a pair on the first roll and then failing to improve it on the subsequent two rolls. Suppose that the first roll consists of a pair and three unmatched dice, an event with probability

$$6 \cdot \binom{5}{2} \cdot \left(\frac{1}{6}\right)^2 \cdot {_5P_3} \cdot \left(\frac{1}{6}\right)^3 = \frac{3600}{7776}.$$

Find the probability that two further rolls of the unmatched dice leave a pair.

On the second roll, it is necessary that the three dice show three different values and fail to duplicate the pair that has been set aside. This probability is

$$\frac{5}{6} \cdot \frac{4}{6} \cdot \frac{3}{6} = \frac{60}{216}.$$

This same probability applies to the third roll, making the probability of completing a pair

$$\left(\frac{60}{216}\right)^2 = \frac{100}{1296}.$$

∎

We see then, that the probability of a pair after 3 rolls is

$$\frac{3600}{7776} \cdot \frac{100}{1296} = \frac{360,000}{10,077,696} \approx .0357.$$

The house edge on a Pair bet paying 25–1 can be shown to be 7.12%, not much less than for a Straight or No Hand.

Example 5.49. A Nutz player may bet on multiple wagers that are covered by a single three-roll sequence. Suppose that a bettor covers each of the 7 wagers with a $1 bet, which will guarantee a win on one of them. What is the probability of a profit?

This wager wins money if the payoff is greater than $7. This occurs when the player rolls a straight, no hand, a pair, or the Nutz. Based on our previous work, we have only to compute the probability of the Nutz. This probability depends on how many rolls are required and how many of the repeated number turn up on each roll.

For convenience, we will first compute the probability of rolling 5 ⚃s. The probability of rolling 5 ⚃s on the first roll and winning the 100–1 payoff is simply

$$\left(\frac{1}{6}\right)^5 = \frac{1}{7776}.$$

Completing the Nutz in two rolls can be broken down into three subcases.

- If 4 ⚃s on the first roll are followed by a ⚃ on the second roll, the probability is

$$\left[\binom{5}{4} \cdot \left(\frac{1}{6}\right)^4 \cdot \frac{5}{6}\right] \cdot \frac{1}{6} = \frac{25}{46,656}.$$

- A first roll with 3 ⚃s without a pair on the other dice followed by 2 ⚃s on the second roll occurs with probability

$$\left[\binom{5}{3} \cdot \left(\frac{1}{6}\right)^3 \cdot \frac{5}{6} \cdot \frac{4}{6}\right] \cdot \left(\frac{1}{6}\right)^2 = \frac{200}{279,936}.$$

- Finally, if 2 ⚃s on the first roll are followed by 3 ⚃s on the second, the probability is

$$\left[\binom{5}{2} \cdot \left(\frac{1}{6}\right)^2 \cdot \frac{5}{6} \cdot \frac{4}{6} \cdot \frac{3}{6}\right] \cdot \left(\frac{1}{6}\right)^3 = \frac{600}{1,679,616}.$$

For a Nutz of 5 ⚃s completed in 3 rolls, there are 6 subcases to be considered. These are listed with their probabilities in Table 5.35, where $x, y,$ and z represent different rolled numbers that are not 4s.

Adding the fourth column gives the probability of rolling 5 ⚃s in exactly 3 rolls: 3.622×10^{-3}. Adding this value to the probabilities for rolling 5 ⚃s in 1 or 2 rolls gives the final probability for the Nutz with ⚃s: 5.358×10^{-3}.

Having computed the probability of rolling 5 ⚃s, we now multiply that by 6 to get .0321: the approximate probability of rolling the Nutz. The probability of making a profit when covering all 7 bets is then

$$.0926 + .0357 + .0321 = .1604,$$

which corresponds to slightly less than one chance in 6. ∎

TABLE 5.35: Nutz: Ways to roll five 4s in exactly 3 rolls

Roll 1	Roll 2	Roll 3	Probability
4444x	4444x	44444	4.465×10^{-4}
444xy	444xy	44444	3.969×10^{-4}
444xy	4444x	44444	1.191×10^{-4}
44xyz	44xyz	44444	9.923×10^{-5}
44xyz	444xy	44444	5.954×10^{-4}
44xyz	4444x	44444	8.931×10^{-4}

Casino Hold 'Em Dice Poker

Another attempt to capitalize on Texas Hold 'Em's rising popularity with a table game replaced the cards with poker dice: special 6-sided dice bearing the faces of playing cards from 9 through ace. Suits, if present, are not typically considered in games played with poker dice. Five poker dice are shown in Figure 5.11.

FIGURE 5.11: Poker dice.

The choice to include "Hold 'Em" in the game's name was more along the lines of a marketing decision than an indicator of how the game is played. There are no community dice in Casino Hold 'Em Dice Poker, nor do players have individual hands which face off against one another. The game begins with each player making an Ante bet, after which a designated player rolls 5 poker dice, which stand as the hand for all players. This hand will be compared to the dealer's complete hand.

The dealer then rolls 2 poker dice, and so the players have some sense of how strong the dealer's hand might be before the decision point. While these two dice give an indication of how strong the dealer's hand might ultimately be, the players' dice have no effect on the dealer's. This is contrary to how card games are played, where the number of aces, for example, in a deck is limited to 4. The players may throw a four-ace hand, but the dealer can still roll 4 (or 5) aces as well.

At this point, players must decide individually whether to fold, surrendering their Ante bet, or make a second Play bet of double the Ante bet. If any player chooses to make the Play bet, the dealer then completes his or her hand by rolling 3 additional poker dice. If the dealer's hand is lower than a pair of kings, he or she fails to qualify, and players who have not folded are paid 1–1 on their Ante bets while their Play bets push.

If the dealer qualifies with a pair of kings or better, the hands are compared. If the players' hand is better, all Ante and Play bets are paid even money; if the dealer's hand is better, the house takes all wagers. An optional bonus of 25–1 for 5 of a kind and 3–1 on 4 of a kind may be paid on the Ante bet.

Hands are ordered as in standard poker, with flushes removed from the ranking:

- 5 of a kind
- 4 of a kind
- Full house
- Straight
- 3 of a kind
- 2 pairs
- Pair
- High card.

However, this order is inconsistent with the probabilities of the various hands when we transition from a deck of cards to a set of poker dice.

Example 5.50. In 5-die poker, a straight is less likely than a full house. The probability of a straight is

$$P(\text{Straight}) = P(9TJQK) + P(TJQKA) = 2P(9TJQK),$$

where T denotes a 10. Continuing, we have

$$P(9TJQK) = \frac{5!}{6^5} = \frac{120}{7776} = \frac{20}{1296},$$

since the 5 card faces may be assigned to the 5 dice in $_5P_5 = 5!$ ways, and the probability of throwing that face on that die is $\frac{1}{6}$. The probability of a straight is thus $\frac{40}{1296}$. The probability of a full house is

$$\frac{6 \cdot \binom{5}{3} \cdot 5}{6^5} = \frac{50}{1296},$$

greater than the chance of a straight. ∎

We next raise the question of the dealer qualifying. We have the following:

$$P(\text{High card hand}) = \frac{80}{1296}, \text{ and}$$

$$P(\text{Pair}) = \frac{600}{1296}.$$

Since the pairs are evenly distributed among the 6 card ranks, the probability that the dealer fails to qualify with at least a pair of kings is

$$\frac{80}{1296} + \frac{400}{1296} = \frac{480}{1296} = \frac{10}{27} \approx .3704.$$

Slightly more than one-third of the time, a hand of Casino Hold 'Em Dice Poker ends early.

When should a player make the Play bet and force the game to a showdown? There are a few principles that should govern this decision independent of any mathematics.

1. If the players' hand is beaten by the dealer's initial two qualifying dice, as when the players hold 99TQA and the dealer shows KK, the players should fold.

2. A weak dealer roll such as 9T might justify playing on a slightly weaker hand than ordinarily in hopes that the dealer might not qualify and a player could win on the Ante bet.

Beyond these simple ideas, a hand should be played if it has a 50% or greater chance of winning, or if it ranks in the top 3888 of the 7776 possible hands. We count up from the bottom.

- 480 hands contain no pair and are not a straight.

- 3600 hands contain a pair, so the cutoff between the top and bottom halves of the 7776 hands contains a high pair. As noted above, pairs are equally distributed among the 6 card ranks, so hand #3888 is a pair of aces. The lowest pair of aces is AAJT9, while the highest is AAKQJ; the line separating the top and bottom halves of all 7776 hands is about two-thirds of the way up the list of ace-pair hands: AAQJT.

Since AAQJT is the lowest-ranked pair containing 2 aces and a 10, one simple rule is to use this as the beacon hand separating folding and playing. Since there are 60 ways to roll AAQJT, which run from 3877–3936 on the ranked list of hands, a more exact rule would be "Fold any hand less than AAK." This would mean playing the top 3876 hands, or the top 49.8%.

As a first approximation, one can do well with the advice "Fold on any hand less than a pair of aces, unless the dealer also has a pair of aces, in which case fold anything less than two pairs." This does not address the case of the dealer rolling a weak initial two cards; we consider that next.

Example 5.51. If the dealer has rolled two cards less than a king, find the probability that the last 3 dice will complete a qualifying hand.

Case 1: The dealer has rolled a pair. The probability of rolling a pair, queens or lower, is $\frac{1}{9}$. The following subsequent rolls will produce a qualifying hand.

- **Another pair.**

  The probability of rolling a second pair not equal to the first pair and producing a two-pair hand (so excluding the possibility of rolling a full house) is

  $$5 \cdot \binom{3}{2} \cdot \left(\frac{1}{6}\right)^2 \cdot \frac{4}{6} = \frac{60}{216}.$$

- **One or more of the card originally rolled.** This includes the possibility of rolling a full house, for example if QQ is followed by Q99.

  $$\left(\frac{1}{6}\right)^3 + \binom{3}{2} \cdot \left(\frac{1}{6}\right)^2 \cdot \frac{5}{6} + \binom{3}{1} \cdot \frac{1}{6} \cdot \left(\frac{5}{6}\right)^2 = \frac{91}{216}.$$

  In this expression, the three terms correspond to final hands of 5 of a kind, 4 of a kind, and either 3 of a kind or a full house with the triple consisting of the rank originally rolled on the first two dice.

- **Three of a kind, but not of the card originally rolled.**

  This roll produces a full house. Its probability is

  $$5 \cdot \left(\frac{1}{6}\right)^3 = \frac{5}{216}.$$

Taken together, the probability of a low pair turning into a qualifying hand is

$$\frac{60 + 91 + 5}{216} = \frac{156}{216}.$$

Case 2: The dealer has not rolled a pair.

There are 6 ways for the dice to fall: QJ, QT, Q9, JT, J9, and T9. The probability of rolling one of these combinations is $\frac{1}{3}$.

- One new possibility here is that the hand completes to a straight. Given any one of the 6 combinations above, there are either 1 or 2 ways to roll three dice to complete a straight. These are shown in Table 5.36.

TABLE 5.36: Casino Hold 'Em Dice Poker: How to roll a straight

Initial roll	Need to roll
QJ	AKT, KT9
QT	AKJ, KJ9
Q9	KJT
JT	AKQ, KQ9
J9	KQT
T9	KQJ

Each way to roll 3d6 and bring up the required 3-card combination has probability

$$_3P_3 \cdot \left(\frac{1}{6}\right)^3 = \frac{1}{36}.$$

Adding across the six starting rolls gives a straight probability of ¼.

- The probability that the last 3 dice produce a pair of kings or aces that is not a two-pair or higher hand is

$$2 \cdot \binom{3}{2} \cdot \left(\frac{1}{6}\right)^2 \cdot \frac{3}{6} = \frac{18}{216}.$$

- If we denote the two card ranks that were initially rolled as X and Y, the probability of completing to a two-pair hand of the form KKXXY or AAXXY is

$$2 \cdot \binom{3}{2} \cdot \left(\frac{1}{6}\right)^2 \cdot \frac{2}{6} = \frac{12}{216}.$$

The other possibility for a two-pair hand is XXYYZ, where Z is different from X and Y. The probability of XXYYZ is

$$3 \cdot \frac{1}{6} \cdot 2 \cdot \frac{1}{6} \cdot \frac{4}{6} = \frac{24}{216}.$$

The total probability of improving XY to two pairs is then

$$\frac{12 + 24}{216} = \frac{1}{6}.$$

$\blacksquare$

5.6   6-Die Games

Six Dice

When the number of dice in a carnival game hits 6, it becomes somewhat more challenging to sort through the possibilities and determine the results, and so such games are rarer than games with 2–4 dice. One very simple proposed game, *Six Dice*, relied only on addition to determine the winner. All payoffs were based on the sum of the 6d6 in use.

Since some of the less common sums have very low probabilities, outcomes were grouped into wagering blocks.

Example 5.52. The probability of rolling a 35 on 6d6, which must consist of five 6s and a 5, is

$$P(35) = \binom{6}{5} \cdot \left(\frac{1}{6}\right)^5 \cdot \frac{1}{6} = \frac{1}{7776}.$$

By grouping 35 with a roll of 8 for betting purposes, the designer establishes a bet with a win probability of $P(8) + P(35)$. An 8 can be rolled on 6 dice as four 1s and two 2s or five 1s and a 3. We have

$$P(8) = \binom{6}{4} \cdot \left(\frac{1}{6}\right)^6 + \binom{6}{5} \cdot \left(\frac{1}{6}\right)^6 = \frac{21}{46,656},$$

and so

$$P(8) + P(35) = \frac{21}{46,656} + \frac{1}{7776} = \frac{1}{1728}.$$

∎

The original patent for Six Dice specified wagers in multiples of 9; this bet paid of at 150–9 [59]. The expected value of a $9 bet on 8 and 35 is

$$E = (150) \cdot \frac{1}{1728} + (-9) \cdot \frac{1727}{1728} = -\frac{15,393}{1728} \approx -\$8.91.$$

The HA on this bet is 98.98%, leaving considerable room to improve the game for players. Paying a winning bet at X to 1 gives an expected return of

$$\frac{X - 1727}{1728};$$

if $X = 1600$, the HA drops to 7.35%. (The original payoff odds correspond to $X = 16\frac{2}{3}$.)

Outcome groups used in this game are shown with their original payoff odds in Table 5.37. The remaining rolls, 19 through 24, were not available for wagering. If the dice showed a sum of 19–24 (Exercise 32), all player bets lost.

Example 5.53. In an effort to encourage wagering on the group containing 6 and 7, the pay table for that group escalated nonlinearly if 27–45 coins were wagered, as shown in Table 5.38. Which of these wagers has the lowest HA?

The probability of rolling a sum of 6 or 7 on 6d6 is

$$p = P(6) + P(7) = \frac{1}{46,656} + \frac{6}{46,656} = \frac{7}{46,656}$$

and the corresponding expectation for a wager of X is

$$E(X,Y) = pY - X(1 - p).$$

Evaluating this expression using Table 5.38 gives HAs from 90.65% for a 45-coin wager up to 95.82% when 18 coins were bet.

While there was some incentive to bet big on the 6–7 combination, the HA was never attractive to the gambler. ∎

TABLE 5.37: Six Dice: Sums of 6d6 used to establish betting groups [59]

Totals	Payoff odds
6,7	3000–9
8,35	150–9
9,10,36	90–9
13,29	85–9
31,32,34	45–9
12,30,33	30–9
11,14,28	20–9
15,16,27	10–9
17,18,25,26	7–9

TABLE 5.38: Six Dice: Payoff odds for increasing bets on 6 and 7 [59]

Wager, X	Payoff, Y
9	3000
18	5000
27	10,000
36	14,000
45	18,000

Turning Six Dice, as written, into something that might be viable in a casino calls for steep increases in the payoffs. Table 5.39 shows an alternate pay table for the 6 and 7 bet, with its corresponding lower HAs.

TABLE 5.39: Alternate pay table for increasing bets on 6 and 7

Wager	Payoff	E	HA
1	5000	–$0.25	24.97%
2	10,000	–$0.50	24.97%
3	17,500	–$0.37	12.46%
4	25,000	–$0.25	6.21%
5	32,500	–$0.12	2.46%

Example 5.54. The wager on 15, 16, and 27 has payoff odds of 10–9, which makes it look like a nearly even proposition. The probability of winning this bet (see Table 5.46 on page 283) is

$$P(15) + P(16) + P(27) = \frac{1666}{46,656} + \frac{2247}{46,656} + \frac{1666}{46,656} = \frac{5579}{46,656} \approx .1196,$$

or just slightly more than $\frac{1}{9}$—nowhere close to $\frac{1}{2}$.

Changing the payoff from 10–9 to 7–1 cuts the HA from 74.76% to a very reasonable 4.34%. ∎

High Roll Dice Match 6

High Roll Dice Match 6 is a 6-die game that provides two simple types of wager. Based on the roll of 6 dice, players may choose from the following bets.

- The "Pip Street" wager is a bet on which number will be rolled most often. Ties are broken by ranking the dice in descending order: ace, 6, 5, 4, 3, 2. A roll of ⚀ ⚁ ⚂ ⚃ pays off Pip Street bets on 4, for example. The pay table, Table 5.40, offers different odds on each number.

TABLE 5.40: High Roll Dice Match 6: Pip Street payoffs

Number	Payoff odds
	8–1
	6–1
	5–1
	4–1
⚄	3–1
⚀	3–1

Example 5.55. What is the probability that 2 will be rolled more than any other number?

The event {2 is rolled most often} can be divided into 6 smaller disjoint events.

A: Two 2s and 4 other distinct numbers.

B: Three 2s and 3 other distinct numbers.

C: Three 2s, a pair, and a sixth die showing neither another 2 or the number in the pair: for example,

D: Four 2s and anything else, including a pair.

E: Five 2s and any 6th number not a 2.

F: Six 2s.

We can compute the probability of each smaller event individually and then add the results. We have

$$P(A) = \binom{6}{2} \cdot \left(\frac{1}{6}\right)^2 \cdot \binom{5}{4} \cdot \left(\frac{1}{6}\right)^4 = \frac{75}{46,656}$$

$$P(B) = \binom{6}{3} \cdot \left(\frac{1}{6}\right)^3 \cdot \binom{5}{3} \cdot \left(\frac{1}{6}\right)^3 = \frac{200}{46,656}$$

$$P(C) = \binom{6}{3} \cdot \left(\frac{1}{6}\right)^3 \cdot 5 \cdot \left(\frac{1}{6}\right)^2 \cdot \frac{4}{6} = \frac{400}{46,656}$$

$$P(D) = \binom{6}{4} \cdot \left(\frac{1}{6}\right)^4 \cdot \left(\frac{5}{6}\right)^2 = \frac{375}{46,656}$$

$$P(E) = \binom{6}{5} \cdot \left(\frac{1}{6}\right)^5 \cdot \frac{5}{6} = \frac{30}{46,656}$$

$$P(F) = \left(\frac{1}{6}\right)^6 = \frac{1}{46,656}$$

and so

$$P(\text{More 2s than any other number}) = \frac{1081}{46,656} \approx .0232.$$

■

- The "Like Kind Boulevard" bet wagers on the number of matches, from 2–6 or 0, of the winning Pip Street number. The 0 bet pays off if there is no one Pip Street winning number: when six different numbers are rolled. Payoff odds are shown in Table 5.41.

TABLE 5.41: High Roll Dice Match 6: Like Kind Boulevard payoffs

Count	Payoff odds
0	60–1
2	1–2
3	2–1
4	15–1
5	200–1
6	6000–1

An expanded version of High Roll Dice Match 6 is played with 10 dice. High Roll Dice is examined beginning on page 276.

5.7   10-Die Games

Twenty-Six

Twenty-Six was a popular game in the American Midwest in the 1950s—in underground gambling operations, of course [88]. To play Twenty-Six, you select a number from 1 to 6—called your *point*—and roll 130 six-sided dice. In practice, this was done by having the player roll ten dice thirteen times. As one might expect from an underground game, the payoff schedule for Twenty-Six varied among establishments. One pay table was this:

- If your point comes up 26–32 times, you are paid off at 3–1 odds.

- If you roll your point 33 or more times, you are paid $7 for every dollar bet.

- At the other end of the scale, if you roll your point 11 or fewer times, that pays 3–1 also.

- Rolling your point exactly 13 times pays 1–1.

The number X of successful rolls of a given point is a binomial random variable with the number of trials $n = 130$ and the probability of success $p = \frac{1}{6}$. The binomial formula gives

$$P(X = x) = \binom{130}{x} \cdot \left(\frac{1}{6}\right)^x \cdot \left(\frac{5}{6}\right)^{130-x}.$$

The mean number of successful rolls is $\mu = np = 21\frac{2}{3}$; reaching any of the payoff thresholds turns out to be a daunting task. Assessing the exact probability of winning a Twenty-Six bet calls for a lot of summation; for example, the chance of rolling 33 or more points is

$$\sum_{x=33}^{130} P(X = x),$$

a sum of 98 terms. These sums are easily computed, but as an experimental investigation of Twenty-Six, 2 million rolls of 130 dice each were simulated with a Python program. A plot of the number X of successful rolls (out of 130) is shown as Figure 5.12.

As an estimate of the probabilities involved, we note that the data form a graph that appears to be bell-shaped and symmetric about its mean μ. This distribution is described by a result called the *Empirical Rule*.

Empirical Rule: If the values of many samples of a random variable with mean μ and standard deviation σ are bell-shaped and symmetric, then we have the following:

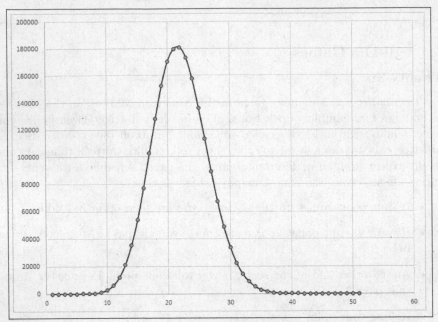

FIGURE 5.12: Number of successes in 2 million simulated games of Twenty-Six.

1. Approximately 68% of the data points lie within 1 standard deviation of the mean: between $\mu - \sigma$ and $\mu + \sigma$.
   This result is often stated as "about 2/3."

2. Approximately 95% of the data points lie within 2 standard deviations of the mean: between $\mu - 2\sigma$ and $\mu + 2\sigma$.

3. Approximately 99.7% of the data points lie within three standard deviations of the mean: between $\mu - 3\sigma$ and $\mu + 3\sigma$.
   This result is often stated as "almost all."

For Twenty-Six, $\mu = 21\frac{2}{3}$ and $\sigma = \sqrt{\dfrac{650}{36}} \approx 4.25$. The Empirical Rule gives the following results:

- Approximately 68% of all 130-die rolls lie in the interval $21\frac{2}{3} \pm 4.25$, or (17.41, 25.92). Since the number of successes in Twenty-Six must be an integer, we round this interval inward (up or down, as needed, to the nearest integer within the interval) to [18, 25]. These results win no money.

- Approximately 95% of the data lies in the rounded interval [14,30]. This interval contains some rolls that pay 3–1, but we are not yet at the 7–1 payoff level, nor have we caught the even money payoff for rolling exactly 13 points.

- The probability that the number of successes lies in the rounded interval [9,34] is approximately .997. This interval contains some of the 7–1 payoffs and also some of the 3–1 payoffs for 11 or fewer successes.

Table 5.42 compares the experimental and theoretical results.

TABLE 5.42: Twenty-Six: Experimental vs. theoretical distribution

Range	Experimental %	Theoretical %
$\mu \pm \sigma : [18, 25]$	63.11%	~ 68%
$\mu \pm 2\sigma : [14, 30]$	94.29%	~ 95%
$\mu \pm 3\sigma : [9, 34]$	99.34%	~ 99.7%

Remembering that the percentages stated in the Empirical Rule are approximate values, we see generally good agreement between experimental results and the theoretical predictions, except perhaps within 1 SD of the mean. The percentage of simulated results within 1 standard deviation of the mean is on the low side; this may be attributable to some inaccuracies introduced in fitting discrete data to a continuous distribution.

These approximations and experimental data suggest that the chance of winning a Twenty-Six bet is small. We can confirm this by using a calculator or computer to find the exact probabilities.

- Rolling exactly 13 points and winning $1 is a simple matter of computing $P(X = 13) \approx .0109$.

- The probability $P(3)$ of winning $3 is

$$P(3) = P(\text{Match 11 or fewer dice}) + P(\text{Match 26–32 dice})$$

$$= \sum_{x=0}^{11} P(X = x) + \sum_{x=26}^{32} P(X = x)$$

$$\approx .1799.$$

- The probability $P(7)$ of winning $7 is

$$P(7) = P(\text{Match 33 or more dice}) = \sum_{x=33}^{130} P(X = x) \approx .0075.$$

The total probability of winning some amount is then about

$$.0109 + .1799 + .0075 = .1982,$$

just under 20%. The chance of losing $1 is approximately .8017.

The house advantage on a round of Twenty-Six is seen to be 19.89%. Despite this high advantage, Twenty-Six operators were tempted to cheat, perhaps by holding out one die and having the player roll only nine, or by miscounting the number of points rolled by a player on a given turn [19].

High Roll Dice

High Roll Dice is a game that offers the exciting possibility of an instant $100,000 payoff on a $1 wager called "the best longshot bet in casino history" [41].

Ten dice are used, and a single game consists of two rolls. On the first roll, known as "Pip Street" as in High Roll Dice Match 6, players may wager on which number is rolled most often. Just like in High Roll Dice Match 6, ties are broken in favor of the highest-ranking dice, with ⚀s highest, followed by ⚄s down to ⚀s. The payoff depends on the winning number, as shown in Table 5.43.

TABLE 5.43: High Roll Dice: Pip Street pay table

Number	Payoff odds
	3–1
	4–1
⚃	4–1
⚃	5–1
⚅	6–1
	6–1

Following the first roll, the winning dice are set aside and the remainder are rolled again. The "Like Kind Boulevard" bet can be made on the number of dice that show the winning number after either the first or second roll, and pays off according to Table 5.44. Players may bet on any number or numbers from 2–8, but must wager an equal amount on each number selected [41].

TABLE 5.44: High Roll Dice: Like Kind Boulevard pay table

Roll	First roll payoff	Second roll payoff
2 of a kind	12–1	N/A
3 of a kind	4–5	6–1
4 of a kind	1–1	1–1
5 of a kind	4–1	2–1
6 of a kind	30–1	4–1
7 of a kind	400–1	10–1
8 of a kind	5000–1	50–1

Example 5.56. Find the probability that the initial roll of the 10 dice yields no more than 2 of any one number, thus triggering a 12–1 payoff on a Like Kind Boulevard bet on 2.

There are 2 distributions that result in a maximum of 2 of any number: 2–2–2–2–2–0 and 2–2–2–2–1–1. The formula for distinguishable permutations (Theorem 1.12, page 21) states that there are

$$\frac{10!}{(2!)^5} = 113,400$$

distinguishable permutations where the dice fall 2–2–2–2–2–0 and

$$\frac{10!}{(2!)^4} = 226,800$$

in the 2–2–2–2–1–1 case. Each of these factors must be multiplied by the number of ways to choose the numbers rolled twice apiece and by $\left(\frac{1}{6}\right)^{10}$, the probability of rolling the specified numbers on their specified dice. The probability is $P(2-2-2-2 \quad 2-0) + P(2-2-2-2-1-1)$, or

$$\frac{\binom{6}{5} \cdot 113,400}{6^{10}} + \frac{\binom{6}{4} \cdot 226,800}{6^{10}} = \frac{175}{2592} \approx .0675.$$

∎

We note that in this instance, the winning number must be ⚀, ⚃, or ⚄. Since there is no additional payoff if the maximum of the chosen number remains 2 after the second roll, the HA on this bet is 12.2%.

The $100,000 payoff is the top prize of a $1 side bet, Millionaire Row, that pays $50,000 if the player rolls 9 of a kind and $100,000 if the 10 dice all show the same number on the first roll. Players need not specify the number on the matching dice, so the probability of winning is

$$P(9) + P(10) = \binom{10}{9} \cdot \left(\frac{1}{6}\right)^8 \cdot \frac{5}{6} + 6 \cdot \left(\frac{1}{6}\right)^{10} = \frac{51}{6^9} \approx 5.0607 \times 10^{-6},$$

slightly more than 1 chance in 200,000.

The expected value of this longshot bet is

$$E = (100,000) \cdot \frac{1}{6^9} + (50,000) \cdot \frac{50}{6^9} + (-1) \cdot \frac{6^9 - 51}{6^9} \approx -\$.7420,$$

so while it may be "the best longshot bet," it's certainly not a good bet.

5.8 Exercises

Solutions begin on page 290.

1-Die Games

1. Find the expected value of a $1 bet on the Shooter's Streak at Die Rich.

2. Beat the Dealer can also be played with one die instead of two. If the house takes all ties, how does the HA of the one-die game compare to the 11.11% edge in two-die Beat the Dealer?

2-Die Games

3. *Under and Over 7* is a simplified two-die game that is sometimes seen at carnivals and charity casino night fundraisers where craps might be considered too complicated or too time-consuming. It can also be played as a street game, with the simple three-square layout, Figure 5.13, drawn with chalk on concrete [88].

Under	7	Over
2,3,4,5,6		8,9,10,11,12
Even money	5 for 1	Even money

FIGURE 5.13: Layout for Under and Over 7.

Three wagers are available: Under 7, 7, and Over 7. Under 7 and Over 7 pay off even money if the sum of two rolled dice is, respectively, less than or greater than 7. The 7 bet pays off 5 for 1 if the next roll is a 7. Find the HA for each wager.

4. Find the house advantage for the first two rolls of a No Lose Free Roll bet.

5. Suppose that the No Lose Free Roll was reconfigured so that any roll of 3, 6, or 9 was a winning roll (the "rollboat"; see [71]). Show that this game results in a player advantage for the $5 bettor.

6. If a No Lose Free Roll bettor is betting $1000 rather than $5, his or her maximum win is capped at $250,000. This allows for only 7 parlays before the cap is reached. Find the house edge of this bet.

7. Which is a better Rocket 7 bet: a $6 wager on Any Double (page 207) or 6 separate $1 wagers on the six individual doubles, which pay 33–1?

8. Construct the PDF for a $1 Survival bet in a 3-roll Survival Dice game after the first two rolls add up to 14.

9. Survival Dice offers an Anything Hard one-roll side bet that pays off at 5 for 1 if the next roll is doubles. Find the HA of this wager.

10. In 3-roll Survival Dice, careful counting with attention to the walls shows that the expected value of a $1 Survival bet is

$$E = -\frac{2861}{46,656} \approx -\$.0613,$$

giving the casino a 6.13% edge. At the Eldorado Casino in Reno, Nevada, the 5000–1 payoff for a sum of 36 was replaced by a progressive jackpot payoff. Find the minimum jackpot amount, rounded to the nearest dollar, for which the Survival bet has positive player expectation.

11. Another wagering option at Heads & Tails was a pass line-like wager on either Heads or Tails, which won if the selected pair was thrown before the other pair [67]. Tosses resulting in 1 head and 1 tail were disregarded and the bet carried over to the next roll. If the payoff on this bet was 9–10, find the HA.

Az'ar

The casino game *Az'ar*, which underwent a field trial at London's Empire Casino in 2016, was unusual in that its rules specified half-inch round-cornered dice instead of standard sharp-edged casino dice [111]. The two types of dice are shown side by side in Figure 5.14. In Az'ar, a selected player shakes two

FIGURE 5.14: Left to right: Standard ¾" casino die and ½" die with rounded edges as used in Az'ar.

dice in a cup, which is then inverted on the table and the dice revealed by the dealer. Dice with rounded corners roll around more freely in a dice cup than perfect cubes, hence the change of dice for Az'ar.

12. A number of wagers are available to Az'ar players; unlike craps, all bets are resolved on every roll. Some Az'ar bets are modifications of craps bets with pushes on certain rolls, which reduces their HAs below the corresponding HAs for craps bets. Find the house advantage on the following Az'ar bets.

a. The *4-5-6* bet pays 3–2 if the next roll totals 4, 5, or 6 and pushes on a roll of 8.

b. The *Hardways* bet pays 7–1 if the next roll is a hard 4, 6, 8, or 10, and pushes on a 2 or 12.

c. *Any 7* pays 4–1, as in craps, if the next roll is a 7, but pushes on any roll totaling 6.

d. Az'ar's *Field* bet pays even money on a 3, 4, 9, 10, or 11; 2–1 on a 2, and 3–1 on a 12. All other rolls lose.

13. Find a roll that should push the following Az'ar bets if the indicated house advantage is desired.

a. *Any craps.* This pays 7–1 on a 2, 3, or 12, and offers a house edge of 5.56%.

b. *8-9-10.* Similar to 4–5–6, the bet pays 3–2 on an 8, 9, or 10. The HA is 2.78%.

c. *Eleven.* The casino pays 15–1 if an 11 is rolled, and holds a 2.78% advantage.

3-Die Games

14. In chuck-a-luck, find an integer payoff for 3 of a kind that gives the smallest possible positive HA.

15. An extra bet seen on some chuck-a-luck layouts is a Field bet, covering the numbers 3–7 and 13–18. This bet pays off at even money. Find the probability of winning and the house advantage of this bet.

16. Consider a hypothetical șic bo bet, "All Low," that pays off if all three dice show a number from 1–3. (An analogous and mathematically identical "All High" bet would cover the case when all three dice come up 4, 5 or 6.) Find the probability of winning this bet and a whole number to 1 payoff that yields the smallest positive HA.

17. In the spirit of the sic bo Triples bet, which pays 30–1 if any triple is rolled, consider a hypothetical *Doubles* bet, which would win if the 3 dice showed two of the same number. It may be desirable to design this bet so that it loses if triples appear; we will consider both possibilities in this exercise.

a. Compute the probability that the three dice show two or more of the same number. (The Complement Rule may be useful here.)

b. Find the house advantage on a $1 Doubles bet with triples excluded and a payoff of 1–1.

c. Repeat part b for a $5 bet if the Doubles bet pays 6–5.

d. Suppose now that the Doubles bet also wins if triples are thrown. Find the HA at payoff odds of 1–1 and 6–5.

18. Another way to accommodate triples in the Doubles bet is to recognize triples as 3 pairs of dice showing the same number, and pay 3–1 for triples while paying even money on a single pair. Find the HA for a $1 bet under these rules.

19. Suppose that the pay table for the yee hah hi Color Double bet was modified to pay 2–1 if 2 dice are the specified color and 3–1 if all three dice are that color. Would this game be better for the gambler than the standard Color Double bet (page 237)?

20. Three Dice Celo offers a confusing sounding wager called "Don't 4 or Better." This pays off at even money if the shooter on a decision roll fails to do at least as well as establishing a point of 4. The bet wins if the shooter rolls a losing combination or sets a point of 2 or 3. Find the probability of winning and the house advantage of this bet.

21. Find the house advantage on Fan Tan's Sheh-Sam-Hong bet.

22. Some Fan Tan games offer an additional wager, the **Gin** bet, which is made on a modified portion of the Fan Tan layout as shown in Figure 5.15.

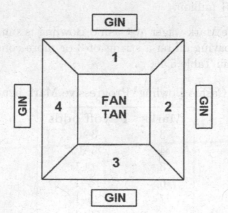

FIGURE 5.15: Alternate Fan Tan layout with Gin bet spaces.

The Gin bet is made on a single number and placed in the rectangle near that number in Figure 5.15. The wager pays even money if that number wins, loses if the number opposite the chosen number wins, and pushes if either of the adjacent numbers wins [39]. Including the 5% commission on winning bets, find the expected return on a $1 Gin bet. How does it compare to the expectation on the similar Nim bet (page 246)?

23. For a bet on two numbers to win at Fan Tan, which has a better return for the player: two separate Fan bets of $1 each, or a single $2 Kwok bet?

24. Rock/paper/scissors could be augmented by an optional Dynamite bet that pays off if the three dice show the same number; this might be viewed by players as a form of insurance. What payoff odds would give a HA between 5% and 10% on this side bet?

4-Die Games

25. Suppose that chuck-a-luck was modified by replacing the 3d6 in the standard cage with 4d8 while maintaining the same payoff structure: x–1 if x dice out of 4 show the chosen number from 1–8.

a. Show that the expected number of matches on a single-number bet is ½.

b. How does the HA of this version compare to the 7.87% edge on all bets at chuck-a-luck played with 3d6?

c. Find the HA if matching all 4 dice pays 100–1 instead of 4–1.

26. Casino Bowling offers a 1-roll Strike bet that pays 5–1 if the next round produces a strike. A 5% commission is charged on winning wagers. Find the expected value and HA.

27. Find the expected value of a $1 Progressive Strike bet if the progressive jackpot is exactly $1 million.

28. The Progressive Mark wager at Casino Bowling is similar to the Progressive Strike wager, paying off on a string of 3 or more consecutive marks. Its pay table is shown in Table 5.45.

TABLE 5.45: Casino Bowling: Progressive Mark bet pay table [8]

Marks	Payoff odds
3	3–1
4	5–1
5	10–1
6	15–1
7	25–1
8	50–1
9	100–1
10	500–1
11	1000–1
12	5000–1

a. Find the probability of rolling n straight marks, for $3 \leqslant n \leqslant 12$.

b. Why is the probability of 12 straight marks higher than the probability of 11 straight marks?

c. Calculate the HA of this wager.

29. In craps, it is customary for a shooter who sevens out to pass the dice to the next player. Casino Bowling could mimic this custom by requiring that a player throwing an open frame yield the dice. Consider a wager that pays 4–1 if three consecutive players toss an open frame and must give up the dice. What is the HA of this wager?

30. If the High and Low bets at Yo! are extended by moving the 11 and 3, respectively, from losing rolls to winning rolls, what is the new HA if the payoff odds remain 3–2?

5-Die Games

31. In Casino Hold 'Em Dice Poker, verify the assertion on page 265 that a high-card hand has probability $\dfrac{80}{1296}$.

6-Die Games

Table 5.46 shows the number of ways to roll sums from 6–36 on 6d6.

TABLE 5.46: Six Dice: Ways to roll sums from 6–36

Sum	Count	Sum	Count	Sum	Count
6	1	17	2856	27	1666
7	6	18	3431	28	1161
8	21	19	3906	29	756
9	56	20	4221	30	456
10	126	21	4332	31	252
11	252	22	4221	32	126
12	456	23	3906	33	56
13	756	24	3431	34	21
14	1161	25	2856	35	6
15	1666	26	2247	36	1
16	2247				

Use this table to solve Exercises 32–35, about Six Dice.

32. Find the probability that a 19–24 is thrown, on which all player bets lose.

33. Consider a $1 bet on the group covering sums of 9, 10, and 36.

3

4 4 Mathematics of Casino Carnival Games

a. Find the probability of winning and express it as a fraction with a numerator of 1.

b. Using the fraction from part a, find a payoff to 1 that is a multiple of 5 (for ease of implementation) and gives a HA between 10% and 12%.

34. For the group bet on 17, 18, 25, and 26, find the probability of winning and the integer payoff odds giving the smallest positive HA.

35. A possible additional pair of Six Dice bets would be *High* and *Low* bets. The High bet wins if the sum of the dice is 22 or higher, while the Low bet wins if the sum is 20 or lower. The most common sum, 21, is a house number on which both High and Low bets lose. Due to symmetry, both bets have the same HA. Assuming an even-money payoff, find this value.

36. One way to turn some rolls of 21 into winning Six Dice rolls would be a *Straight* bet, which would pay off if every die showed a different number.

a. Find the probability of rolling a 6-number straight.

b. If the Straight bet pays 60–1, find the casino's edge.

Appendix A

Answers to Selected Exercises

Answers are included here for problems calling for a numerical solution. Solutions to problems that ask for verification of a formula are not listed.

Chapter 2

Exercises begin on page 46.

1a. 13.99%.

1b. –6.42%—the player would have the edge.

2. No.

3. 6.

4a. A 4–3 payoff gives a 0.08% HA.

4b. A 4–3 payoff gives a 4.40% HA.

4c. A 9–4 payoff gives a 0.69% HA.

5a. $E(x) = \dfrac{x - 15}{16}; x = 14.$

5b. $E(x) = \dfrac{x - 7}{8}; x = 6.$

5c. $x = 15; x = 7.$

6. 60–1 or 61–1.

7.

Value	Prob.
5	6/64
6	10/64
7	12/64
8	12/64
9	10/64
10	6/64

8a. $9\frac{2}{3} - 1.$

8b. $19\frac{1}{3} - 1.$

9. 12.5%.

10a. $E(x) = \dfrac{x-35}{16}; x = 34.$

10b. $E(x) = \dfrac{x-17}{18}; x = 16.$

10c. $x = 35; x = 17.$

11. $\frac{13}{108} \approx .1204.$

12. $\frac{15}{216} \approx .0694.$

13. .1336.

14. 6–1.

15. 5.50%—see [60].

Chapter 3

Exercises begin on page 77.

1a. $\dfrac{72-6n}{n}$, rounded to the nearest cent.

1c. $\sigma^2 = 3n^2 - 39n + 468.$

1d. The standard deviation is $\sigma = \sqrt{3n^2 - 39n + 468}.$

2. Variances are multiplied by either 9 or 25; SDs are multiplied by 3 or 5, respectively.

4.

Number of cards	New HA
1	8.93%
2	7.14%
4	7.14%
8	0.00%
13	–16.07%

5. 7.14%.

6. 3.70%.

7b. South Australia and Western Australia.

7c. New South Wales and Queensland.

8. The HA on the blue Line bet is 25%; the closest HA to this is the yellow Line bet, at 20%. (Other HAs computed in the text all fall below 25%.)

9.

Wager	Payoff odds
Single-number or white pig	40 for 1
Pink or blue pigs	20 for 1
Line	5 for 1
Odd/Even	2 for 1

10. 20 for 1 gives a HA of 11.11%.

11. 8 for 1.

12. Flush is better than Open Straight.

13. Flush is better for the player than 3 of a kind, and has the same HA as a bet on the 5 rank.

14. .0792.

15a. $D_7 = 1854; D_8 = 14,833$.

15b. Possibly the easiest way to do this is to extend the three barred permutations from 6-ball Rouleno by adding on the 7 and 8 in order, and then following the pattern from those three permutations to a similar permutation with the 7 and 8 reversed and all other numbers in their correct places. The resulting 4 permutations would be 1 2 3 4 6 5 7 8, 1 2 4 3 5 6 7 8, 2 1 3 4 6 5 7 8, and 1 2 3 4 5 6 8 7.

15c. .1249. Paying 7–1 gives a HA of 0.06%.

15d.

x	$P(X = x)$
0	0.3679
1	0.3679
2	0.1840
3	0.0611
4	0.0156
5	0.0028
6	0.0007
7	0
8	2.4802×10^{-5}

15e. .3686.

15f. .0191. Paying 51–1 gives a HA of 0.44%. (Paying 50–1 would be easier on dealers and would give a 2.35% HA.)

16. 14.27%.

288   *Mathematics of Casino Carnival Games*

Chapter 4

Exercises begin on page 189.

1. Four-of-a-kinds: 14,464. Straight flushes: 47,752.

2. 84.

3. 1.54%, 1.61%, 1.69%.

4a. $P(\text{Flush}) = \dfrac{11n - 1}{52n - 1}.$

4b. 0.39%, 0.32%., 0.24%.

5. $\frac{4}{221} \approx .0181.$

6a. Fold.

6b. Fold. ($E = -\$.375$.)

7. $\dfrac{7}{47}.$

8. .9234¢.

9. The HA on Any 5 Red and Any 5 Black is the slightly better 14.21%.

10. $p \approx .0286.$

		Probability	Expectation	HA
	a.	0.0020	-0.6038	60.38%
11:	b.	0.0164	-0.4767	47.67%
	c.	0.0253	-0.2407	24.07%
	d.	0.0003	-0.8781	87.81%

12. No. The new HA is 13.46%.

13a. $2.316 \times 10^{-5}.$

13b. .0073.

13c. .0658.

14. 7.95%.

15a. Royal flush.

15b. $8.142 \times 10^{-5}.$

15c. Yes.

16a. .1498.

16b. .1055.

18a. $4.809 \times 10^{-4}.$

18b. 89.

20. The expectation of $1 bet with the infinite-deck approximation is 6.25%.

21. The worst hand, with a score of 5, includes 3 sets of a suited 2 and 3, and a single card—the 2, 3, 4, or 5—in the fourth suit. There are 16 such hands; the probability of being dealt a hand scoring 5 is $\dfrac{16}{\binom{52}{7}} \approx 1.196 \times 10^{-7}.$

22a. $3.232 \times 10^{-5}.$

22b. .0013.

23a. $8.423 \times 10^{-5}.$

	Outcome	Probability
23b.	Win	.2564
	Lose	.2564
	Push	.2991
	Split	.1880

24. .0376.

25. This question asks for the probability that all 4 cards of the chosen rank fall into even-numbered positions in the deck, except for positions 50 and 52, which are revealed only under calling the turn. This leaves 24 possible even-numbered positions, and so the probability is

$$\frac{\binom{24}{4}}{\binom{52}{4}} \approx .0393.$$

26a. $P(\text{Match } k \text{ cards}) = \dfrac{\binom{12}{k} \cdot \binom{60}{3-k}}{\binom{72}{3}}.$

26b.
$$\sum_{k=0}^{3} k \cdot P(\text{Match } k \text{ cards}) = \frac{1}{2}.$$

26c. −\$.0738.
26d. 7.01%.

	Decks	P(3 of a kind)
	1	.0024
	2	.0040
	3	.0046
27.	4	.0049
	5	.0051
	6	.0052
	7	.0053
	8	.0054

28. 1. Connector: ⚃⚁. Hand: ⚀.
2. Connector: ⚁. Hand: , ⚄, ⚄.
29. 20.
30. 29.12%.
31a. .1548.
31b. .4376.
32a. 2.124×10^{-4}.
32b. 4.248×10^{-4}.
32c. .0102.

33. The dealer must hold 6 face cards, since any set of 6 of the remaining 18 face cards will outrank 6 jacks. The probability is

$$\frac{\binom{18}{6}}{\binom{46}{6}} \approx .0020.$$

34a. $\dfrac{1}{4096}$.

34c. Answers will vary. A pay table of 3–1 on 2 matches, 60–1 on 3 matches, and 1000 on 4 matches gives a 5.15% HA.

34d. Answers will vary. A pay table of 1–1 on 2 matches, 10–1 on 3 matches, 500–1 on 4 matches, and 2000–1 on 4 matches gives a 3.00% HA.

Chapter 5

Exercises begin on page 277.

1. 16.52%.

2. With one die, the HA is 16.67%, so this bet is worse than in the 2-die game.

3. The HA on all 3 bets is 16.67%.

4. 29.01%.

5. The player edge with this win probability is 8.91%.

6. 10.31%.

7. Neither one.

8.

Payoff	Prob.
−1	15/36
0	18/36
1	2/36
2	1/36

9. 16.67%.

10. $7862.

11. 5%.

12a. 2.78%.

12b. 5.56%.

12c. 2.78%.

12d. 2.78%.

13a. 4 or 10.

13b. 6.

13c. 4 or 10.

14. 34–1.

15. $P(\text{Win}) = \dfrac{91}{216}$. HA = 15.74%.

16. $P(\text{Win}) = \dfrac{1}{8}$; 7–1.

17a. $\dfrac{96}{216} = \dfrac{4}{9}$.

17b. $16\frac{2}{3}\%$.

17c. $8\frac{1}{3}\%$.

17d. 11.11%, 2.22%.

18. 5.56%.

19. No. The HA of this bet is 18.51%. Color Double carries a HA of 11.11%.

20. $\dfrac{51}{108}$; 5.56%.

21. 1.25%.

22. 1.25%—the same as for Nim.

23. Fan carries a 3.75%. HA. Kwok's HA is 2.50%.

24. At 32–1, the HA is 8.33%, and if the bet pays 33–1, the HA drops to 5.56%.

25b. 8.62%.

25c. 6.27%.

26. $-4\frac{1}{6}$¢; 4.17%.

27. 1.83%.

28a. If $p = P(\text{Mark}) = \dfrac{565}{1296}$, then $P(n \text{ marks}) = p^n(1 - p)$.

28b. Since the game stops after 12 frames, $P(12 \text{ marks}) = p^{12}$ also covers the very small probability of rolling more than 12 marks.

28c. 5.05%.

29. 10.28%.

30. 2.78%.

32. $\dfrac{24,017}{46,656} \approx .5148$.

33a. $\dfrac{183}{46,656} \approx \dfrac{1}{255}$.

33b. At 225–1, the HA is 11.46%.

34. $P(\text{Win}) = \dfrac{11,390}{46,656}$. At 3–1, the HA is 2.35%.

35. 9.28%.

36a. $\dfrac{_6P_6}{6^6} = \dfrac{720}{46,656} \approx .0154$.

36b. 5.86%.

References

[1] Ancient Chinese Games Popular, *Reno Gazette-Journal*, 22 November 1980, p. 20.

[2] Asbury, Herbert, *Sucker's Progress: An Informal History of Gambling in America*. Thunder's Mouth Press, New York, 1938.

[3] Bernard, Jean-Michel, Tapis Vert—Carré d'as. Online at `https://www.youtube.com/watch?v=ehGguw1Daa8`. Accessed 30 May 2018.

[4] Bartle, Richard J.E., Method of Playing a Three Dice Betting Game. US Patent #5,308,081. Online at `https://patents.google.com/patent/US5308081A`. Accessed 8 August 2019.

[5] Braun, Julian, Braun Answers. *Systems & Methods* 9, pp. 52–57, 1975.

[6] Bollman, Mark, *Basic Gambling Mathematics: The Numbers behind the Neon*. CRC Press/Taylor & Francis, Boca Raton, FL, 2014.

[7] Bollman, Mark, *Mathematics of Keno and Lotteries*. CRC Press/Taylor & Francis, Boca Raton, FL, 2018.

[8] Boylan, Eugene B., *et al.*, Method of Playing a Dice Wagering Game Simulating Bowling. United States Patent #5,505,457. Online at `https://patents.google.com/patent/US5505457`. Accessed 15 September 2019.

[9] Callahan, Bryan, New game making its way to Nevada casinos. KTNV-TV, Las Vegas, NV, 3 October 2018. Online at `https://www.ktnv.com/news/new-game-making-its-way-to-nevada-casinos`. Accessed 4 October 2018.

[10] Card-o-lette. *New Jersey Register* 25 N.J.R. 2231, #11, 7 June 1993. Online at `https://www.njstatelib.org/wp-content/uploads/law_files/imported/Research_Guides/Law/njregister/volume25number11%20page2151-2620.pdf`. Accessed 8 February 2020.

[11] Casino Cribbage rack card. Online at `https://wizardofodds.com/blog/raving-2014/images/Cribbage001.jpg`. Accessed 7 June 2018.

[12] Casino Dice. Online at http://www.latestbingobonuses.com/free-games/casino/casino-dice.html. Accessed 3 September 2015.

[13] Casino Dice. Online at `http://www.beatingbonuses.com/casinodice.htm`. Accessed 3 September 2015.

[14] Craps Game with a Repeated Number Based Wagering Area, United States Patent Application Publications, # US 2014/0138911 A1, 22 May 2014. Online at `https://patents.google.com/patent/US20140138911`. Accessed 7 April 2018.

[15] de Moivre, Abraham, *The Doctrine of Chances*, reprint of 3rd (1756) edition. New York, Chelsea Publishing Company, 1967.

[16] Dibattista, Laurie, Playing to Win: Entrepreneur Bets on "Pyramid Dice." *Denver Business Journal*, 9 August 1998. Online at `https://www.bizjournals.com/denver/stories/1998/08/10/smallb1.html`.

[17] Dice Duel. Online at `https://www.crownmelbourne.com.au/casino/casino-games/dice-duel`. Accessed 27 April 2018.

[18] DiLullo, Dean, Betting Game Method of Play, United States Patent #5,350,175, 27 September 1994. Online at `https://patents.google.com/patent/US5350175A/`. Accessed 14 July 2019.

[19] Drzazga, John, Gambling and the Law—Dice, 43 *Journal of Criminal Law & Criminology*, **43** #3, pp. 405–411, 1952.

[20] Empire Global Gaming, Inc. Launches a New 50 Card Deck. Online at `http://empireglobalgaminginc.com/empire-global-gaming-inc-launches-a-new-50-card-deck/`. Accessed 3 May 2015.

[21] Epstein, Richard A., *The Theory of Gambling and Statistical Logic*, 2nd edition. Oxford, Academic Press, 2013.

[22] Ethier, S. N., Faro: From Soda to Hock, in *Optimal Play: Mathematical Studies of Games and Gambling*, Stewart N. Ethier and William R. Eadington, editors. Institute for the Study of Gambling and Commercial Gaming, Reno, NV, 2007.

[23] Euler, Leonhard, On the Advantage to the Banker in the Game of Pharaon. *Mémoires de l'académie des sciences de Berlin* [20] (1764), 1766, pp. 14–164. Online at `https://www.cs.xu.edu/math/Sources/Euler/E313.pdf`. Accessed 7 July 2018.

[24] Fan-Tan-Astic Dice. Online at `http://xcitegaming.com/fantan.html`. Accessed 13 September 2015.

[25] Film8ker, American Roulette Table Layout.gif. Online at http://commons.wikimedia.org/wiki/File%3AAmerican_roulette_table_layout.gif. Accessed 16 May 2013.

[26] A First Look at the Long-Awaited Street Dice at Downtown Grand Las Vegas. Online at http://vitalvegas.com/first-look-long-awaited-street-dice-downtown-grand-las-vegas/. Accessed 19 May 2014.

[27] Frey, Richard L., *According to Hoyle*. Fawcett Crest Books, Greenwich, CT, 1970.

[28] Game Rules Rocket 7. Online at `https://docslide.com.br/ documents/game-rules-rocket-7-swiss-casinos-zuerich.html`. Accessed 5 June 2018.

[29] Gaming Studio, Inc. Online at `https://www.gamingstudio.com/`. Accessed 15 February 2020.

[30] Goott, Joseph, Poker-Keno Game. United States Patent #4,364,567, 21 December 1982. Online at patents.google.com.patent.US4364567. Accessed 16 July 2018.

[31] Gordon, Michael P., An analysis of the game of Pell with a suggested strategy to win. *Casino & Sports* **18**, 73–77.

[32] Gray, Patrick, Board Game. United States Patent #4,635,938, 13 January 1987. Online at patents.google.com.patent.US4635938A. Accessed 23 August 2019.

[33] Grebert-Craig, Brooke, Two-up: How to play and why it is illegal except on Anzac Day. *The New Daily*, 27 April 2018. Online at `https://thenewdaily.com.au/news/national/2018/04/ 27/two-up-game-anzac-day/`. Accessed 18 May 2018.

[34] Griffin, Gavin, First, a Beginner. *Card Player* **31** #22, 24 October 2018, pp. 37–39.

[35] Griffin, Peter, and John M. Gwynn, Jr., An Analysis of Caribbean Stud Poker, in *Finding the Edge*, Olaf Vancura, Judy A. Cornelius, and William R. Eadington, editors. Institute for the Study of Gambling and Commercial Gaming, Reno, NV, 2000, pp. 273–284.

[36] Griffin, Tina, Letter to Ashford Kneitel, Ashford Gaming LLC, 24 April 2014. Online at `https://fortress.wa.gov/wsgc/etransfer/ OnlineServices/activities/ViewGameRule.cshtml?gid=9`. Accessed 2 March 2019.

[37] Haigh, John, *Taking Chances: Winning with Probability*, 2nd edition. Oxford University Press, Oxford, 2003.

[38] Hall, Geoffrey W., Methods of Administering a Wagering Game, US Patent Application Publication #US 2016/0016070 A1. Online at `https://patents.google.com/patent/US20160016070`. Accessed 22 December 2019.

[39] Helprin, Syd, *The Gambling Times Guide to European and Asian Games*. Gambling Times, Inc., Hollywood, CA, 1986.

[40] Helprin, Syd E., Raking in the Ringgits. *Gambling Times* 4 #2, May 1980, pp. 62-78.

[41] High Roll Dice. Online at `https://www.gigaming.net/high-roll-dice/`. Accessed 13 February 2020.

[42] How, Stephen, Discount Gambling: High Card Flush. Online at `https://discountgambling.net/2013/04/22/high-card-flush/`. Accessed 13 June 2018.

[43] Imaginamics, LLC, Introducing Flip-It. Online at `https://imaginamics.net/introducing-flip-it/`. Accessed 29 May 2018.

[44] Jacobson, Eliot, *Advanced Advantage Play*. Blue Point Books, Santa Barbara, CA, 2015.

[45] Jacobson, Eliot, *Contemporary Casino Table Game Design*. Blue Point Books, Santa Barbara, CA, 2010.

[46] Kaplan, Michael, Casino Killers: How a Harvard Maths Graduate is Beating Vegas. *Wired*, January 2011. Online at `http://www.wired.co.uk/magazine/archive/2011/01/features/casino-killers?page=all`. Accessed 1 August 2014.

[47] Kal Gaming International, Pyramid Dice. Online at `http://www.kalgaming.com/pyramiddice.htm`. Accessed 25 April 2018.

[48] Lichtenstein, Sarah, and Paul Slovic, Response-Induced Reversals of Preference in Gambling: An Extended Replication in Las Vegas. *Journal of Experimental Psychology* 101, #1, 16–20, 1973.

[49] Live Casino Comparer, Evolution Dreamcatcher. Online at https://www.livecasinocomparer.com/live-casino-software/evolution-live-casino-software/evolution-dreamcatcher/. Accessed 1 March 2018.

[50] Lubin, Dan, *The Essentials of Casino Game Design*. Huntington Press, Las Vegas, 2016.

[51] Lunar Poker—How to Play. Online at `http://www.lunarpoker.com/how-to-play-lunar-poker/`. Accessed 18 August 2018.

[52] Maguire, Michael, Yee Hah Hi. Online at http://www.australian gambling.com.au/online-casinos/sic-bo/yee-hah-hi/. Accessed 2 October 2015.

[53] Massachusetts Gaming Commission, Two Card Joker Poker. Online at https://massgaming.com/wp-content/uploads/Two-Card-Joker-Poker.pdf. Accessed 18 June 2018.

[54] Meyer, Don, The Game of Faro. Online at http://www.dpmeyer.com/pdfs/Faro.pdf. Accessed 2 June 2018.

[55] Mickey Mouse Roulette Snubbed by Gamblers Here, *Nevada State Journal*, 15 May 1938, p. 1.

[56] Mickey Mouse Wheel to Be Tried on Reno, *Nevada State Journal*, 11 May 1938, p. 1.

[57] Midnight, Ron, French Cafe Gambling Society. *WIN Magazine* **13** #3, April 1991, pp. 46–48.

[58] Montana Lottery, $hake a Day. Online at http://montanalottery.com/treasurePlay. Accessed 11 May 2016.

[59] Moody, Ernest W., Six Dice Game. US Patent #6,854,732 B2, 15 February 2005. Online at https://patents.google.com/patent/US006854732B2. Accessed 19 October 2019.

[60] Mourad, Raphael, Wagering Game Using Dice or Electronically Simulated Dice. US Patent Application Publication #US2009/0186688 A1, 23 July 2009. Online at https://patents.google.com/patent/US20090186688A1/en. Accessed 11 October 2019.

[61] "Mouse Roulette" Game Fails in Reno, *Oakland Tribune*, 16 May 1936.

[62] Mystery Card Bonanza—Procedures. Online at https://www.inag11.com/copy-of-roulette-style-games. Accessed 7 September 2019.

[63] Nazelrod, Scott, *The Denexa Book of Card Games*, version 3. Denexa Games LLC, Norman, OK, 2017.

[64] Nevada Gaming Commission Approved Gambling Games Effective April 1, 2018. Online at http://gaming.nv.gov/Modules/ShowDocument.aspx?documentid=7097. Accessed 20 April 2018.

[65] Nunn, Angela, Popular "Shake-a-Day" Game Misunderstood. Montana Gaming Group, 1 April 2012. Online at http://www.montanagaminggroup.com/index.aspx/Regulation/Popular_shakeaday_game_mis understood. Accessed 11 May 2016.

[66] Nutz. Online at https://www.crownmelbourne.com.au/casino/casino-games/nutz. Accessed 27 April 2018.

[67] Ollington, Robert F., Casino Table Game and Dice. US Patent #4,688,803, 25 August 1987. Online at https://patents.google.com/patent/US4688803A. Accessed 26 September 2019.

[68] The Origins and Anatomy of a New Game: Playing Pell at Sam's Town. *Casino & Sports* **18**, 68–72.

[69] O'Sullivan, Stephen J., Now There's Red Dog! And Here's How It's Played. *Gambling Times* **11** #6, January/February 1988, p, 44–47.

[70] Page, Robert A., Casino Dice Game. US Patent #5,133,559, 28 July 1992. Online at `https://patents.google.com/patent/US5133559A`. Accessed 4 August 2019.

[71] Paulsen, Dennis, A Theoretical & Computer Analysis of Vegas World's Innovative 'No Lose Free Roll' Craps. *Casino & Sports* **26**, 71–78.

[72] Poker All: The New Game in Town, *Gambling Times* **4** #2, May 1980, pp. 80–81.

[73] Poquet, Pierre, File:Boule01.jpg. Online at `https://commons.wikimedia.org/wiki/File:Boule01.jpg`. Accessed 24 July 2018.

[74] Poundstone, William, *Priceless: The Myth of Fair Value (and How to Take Advantage of It)*. Hill and Wang, New York, 2010.

[75] Pyramid Dice Promo (video). Online at `https://www.youtube.com/watch?v=3YJc9WIwAck`. Accessed 25 April 2018.

[76] Quinn, John Philip, *Fools of Fortune or Gambling and Gamblers*. G.L. Howe and Company, Chicago, 1890. Online at `http://www.gutenberg.org/files/58280/58280-h/58280-h.htm`. Accessed 16 January 2019.

[77] Racing Card Derby. Online at `http://racingcardderby.com/products/racing-card-derby-table-game/`. Accessed 11 September 2015.

[78] Ray, Randy, Mississippi Stud Strategy—How to Play and Win this Poker Game, 1 August 2017. Online at `https://www.gamblingsites.com/blog/mississippi-stud-strategy-play-and-win-36420/`. Accessed 15 August 2018.

[79] Richardson, Joe, Tri-Wheel® circa 1988, 25 June 2007, Online at `http://minnesotatri-wheel.blogspot.com/2007/06/tri-wheel-circa-1988.html`. Accessed 15 February 2020.

[80] Richardson, Joe, How Is It Played?, 5 July 2007, Online at `http://thepigwheel.blogspot.com/2007/07/how-it-is-played.html`. Accessed 15 February 2020.

[81] Rio Suite has "'Survival Dice", *Los Angeles Times*, Main edition, 11 September 1994, p. 405.

[82] Roulette Table Layouts. Online at `http://www.ildado.com/roulette_table_layout.html`. Accessed 18 June 2012.

[83] Royer, Victor M., *Powerful Profits from Casino Table Games*. Kensington Publishing Corp., New York, 2004.

[84] Rules for Federal Wheel. Online at `http://www.treasury.tas.gov.au/...Federal_Wheel.../Gaming_Rules_Federal_Wheel.pdf`. Accessed 5 November 2015.

[85] Rules of the Games. New Jersey Division of Gaming Enforcement, 7 November 2011. Online at `https://www.state.nj.us/oag/ge/docs/ProposedRules/jan62012/69F.pdf`. Accessed 8 February 2020.

[86] Scarne, John, *The Amazing World of John Scarne*. Crown Publishers, Inc., New York, 1956.

[87] Scarne, John, *Scarne on Card Games*. Dover Publications, Inc., Mineola, NY, 2004 reprint of 1965 revision.

[88] Scarne, John, *Scarne on Dice*, 8th revised edition. Crown Publishers, Inc., New York, 1980.

[89] Scarne, John, *Scarne's Complete Guide to Gambling*. Simon and Schuster, New York, 1961.

[90] Schlesinger, Don, The Gospel According to Don... *Blackjack Forum* **XI** #3, September 1991, pp. 19–21.

[91] Shackleford, Michael, 3 Card Blitz. Online at `https://wizardofodds.com/games/3-card-blitz/`. Accessed 29 March 2020.

[92] Shackleford, Michael, DJ Wild. Online at `https://wizardofodds.com/games/dj-wild/`. Accessed 2 August 2019.

[93] Shackleford, Michael, Lunar Poker. Online at `https://wizardofodds.com/games/lunar-poker/`. Accessed 18 August 2018.

[94] Shackleford, Michael, Red Dragon. Online at `https://wizardofodds.com/games/red-dragon/`. Accessed 3 November 2019.

[95] Shackleford, Michael, Three Card Poker. Online at `http://wizardofodds.com/games/three-card-poker/`. Accessed 19 July 2012.

[96] Shackleford, Michael, Triple Shot. Online at `https://wizardofodds.com/games/triple-shot/`. Accessed 15 June 2018.

[97] Shackleford, Michael, Wild Five Poker. Online at `https://wizardofodds.com/games/poker-with-a-joker/`. Accessed 7 March 2019.

[98] Sheinwold, Alfred, 4–5–6: The Advantage Is Rolling First. *Gambling Times* **6** #12, February 1984, p. 13.

[99] Sic bo table.png. Online at `http://en.wikipedia.org/wiki/File:Sic_bo_Table.png`. Accessed 16 May 2013.

[100] Simelum, Maki Stanley, Republic of Vanuatu Casino Control Act [CAP 223], Declaration of Authorised Game (Lunar Poker) Order No. 170 of 2013. 267 December 2013. Online at `http://www.paclii.org/vu/legis/sub_leg/ccadoagpo2013584.pdf`. Accessed 18 August 2018.

[101] Sklansky, David, and Alan E. Schoonmaker, *DUCY? Exploits, Advice, and Ideas of the Renowned Strategist.* Two Plus Two Publishing LLC, Henderson, NV, 2010.

[102] Sklansky, David, Let It Ride: A New Casino Game. *Gambling Times* **2** #2, April 1978, pp. 72–73.

[103] Snyder, Arnold, Will Pell Sell? *Blackjack Forum* **2**, 16–18. June 1982.

[104] Stearns, John, Casinos are going to war. *Reno Gazette-Journal,* 11 May 1994. Online at `https://www.newspapers.com/newspage/154095342/`. Accessed 20 June 2018.

[105] Super Color Sic Bo. Online at `http://royal369info.com/en/super-color-sic-bo/`. Accessed 13 April 2018.

[106] Tai Sai. Online at `https://www.skycityauckland.co.nz/casino/table-games/tai-sai/`. Accessed 2 October 2015.

[107] Thorp, Edward O., Nonrandom Shuffling with Applications to the Game of Faro. *Journal of the American Statistical Association* **68** #344, December 1973, pp. 842–847.

[108] Toplikar, Dave, Once king of the gambling halls, faro now a ghost. *Las Vegas Sun*, 19 November 2012. Online at `https://lasvegassun.com/news/2012/nov/19/once-king-gambling-halls-faro-now-ghost/`. Accessed 3 June 2018.

[109] Turning Stone Resort Casino Table Games. Online at `http://www.turningstone.com/gaming/table-games`. Accessed 2 June 2014.

[110] Ubetcha Games: Top Rung Game Play. Online at `http://www.ubetchagames.com/top_rung/gamePlay.html`. Accessed 31 July 2014.

[111] UK Casino Table Games: Az'ar. Online at `http://www.ukcasinotablegames.info/diceazar.html`. Accessed 28 August 2018.

[112] UK Casino Table Games: Diceball. Online at `http://www.ukcasinotablegames.info/diceball.html`. Accessed 7 September 2018.

[113] UK Casino Table Games: Duelling for Dollars. Online at `https://www.ukcasinotablegames.info/duellingdollars.html`. Accessed 22 February 2020.

[114] UK Casino Table Games: Qilin—(Basic Game). Online at `http://www.ukcasinotablegames.info/qilin.html`. Accessed 18 April 2018.

[115] UNLV Online Gaming Abstract: Atlantic City, New Jersey 1992. Center for Gaming Research, University of Nevada, Las Vegas. Online at `https://gaming.unlv.edu/abstract/ac1992.html`. Accessed 7 February 2020.

[116] UNLV Online Gaming Abstract: Atlantic City, New Jersey 1993. Center for Gaming Research, University of Nevada, Las Vegas. Online at `https://gaming.unlv.edu/abstract/ac1993.html`. Accessed 7 February 2020.

[117] Winnovations: Pokette. *WIN Magazine* **14**, #3, n.d.

[118] Yandek, Christopher, Las Vegas Game Series: Faro and Faro Casino Chips. Online at https://spinettisgaming.com/blogs/casino-gaming-history-news/las-vegas-game-series-faro-and-faro-casino-chips. Accessed 2 June 2018.

[119] Yo! Game Play Description. Online at `https://www.playmoreyo.com/game-description/`. Accessed 9 November 2018.

Index

Printed in the United States
by Baker & Taylor Publisher Services

Printed in the United States
by Baker & Taylor Publisher Services